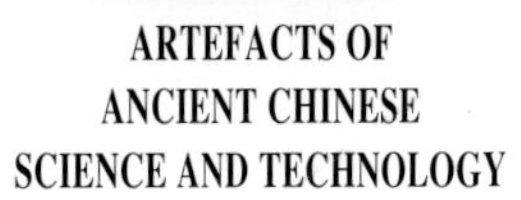

ARTEFACTS OF
ANCIENT CHINESE
SCIENCE AND TECHNOLOGY

Managing Editor: Jiang Cheng'an

Photographs: Yan Zhongyi

Translated into English: He Fei

Design: Zheng Hong

ARTEFACTS OF ANCIENT CHINESE SCIENCE AND TECHNOLOGY

COMPILED BY THE EDITORIAL BOARD
OF THE ARTEFACTS OF ANCIENT CHINESE
SCIENCE AND TECHNOLOGY

Published by:
MORNING GLORY PUBLISHERS
35 Chegongzhuang Xilu Beijing 100044 China

Distributed by:
CHINA INTERNATIONAL BOOK TRADING CORPORATION
35 Chegongzhuang Xilu Beijing 100044 China
(P.O.Box 399, Beijing, China)

First Edition First Printing 1998
ISBN 7-5054-0565-9/J·0287
08600

Printed in the People's Republic of China

ARTEFACTS OF ANCIENT CHINESE SCIENCE AND TECHNOLOGY

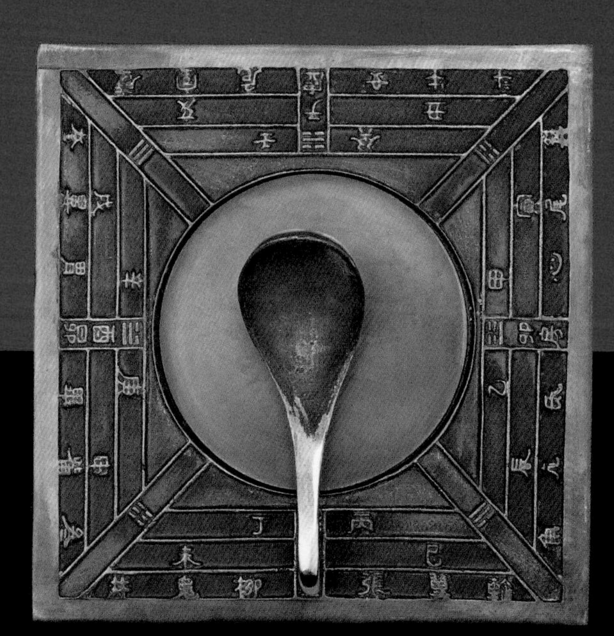

CONTENTS

FOREWORD

Natural science and engineering technology came into existence and grew to meet the needs of production and life of mankind. Starting from imitating, utilising and transforming nature, man made different kinds of substances and installations which had not existed in the natural world for the sake of production and life. The history of science and technology has formed an intimate and inseparable part of the civilisation of society, playing a crucial role in man's progress.

As one of the world's four major countries with ancient civilisation, China blazes gorgeous splendour in natural science and engineering technology. She made outstanding achievements in ancient astronomy, mathematics, physics, chemistry, geology, biology, agriculture, medical and pharmaceutical science, building construction, textiles, ceramics, shipping, navigation and water conservancy. The achievements are a summing up of Chinese forebear's knowledge and adaptation of the laws of nature. It may be said that they are the rewards to Chinese ancestors in changing environment and winning broader space for existence under permissible conditions. In a word, the achievements are a crystallization of the industry and wisdom of thousands of labourers. Based on the inventions and discoveries of obscure labourers, many great scientists and skilful craftsmen have emerged in the course of the development of ancient science and technology in China, such as: Lu Ban, Bian Que, Zhang Heng, Cai Lun, Zhang Zhongjing, Hua Tuo, Ma Jun, Zu Chongzhi, Sun Simiao, Yi Xing, Bi Sheng, Shen Kuo, Li Jie, Yang Hui, Guo Shoujing, Zhu Shijie and Li Shizhen, who are outstanding representatives.

Performing remarkable feats they revealed the mystery of nature and brought happiness to man, greatly enriching the contents of Chinese achievements in science and technology. Their names are revered by posterity forever.

The achievements of ancient Chinese science and technology are the common wealth of the whole of mankind. They belong to China and the world. In the course of its development China has absorbed the fine fruits of labour of many nations and regions. At the same time through various channels China has offered the fruits of her labour to mankind. Her rich data of ancient astronomical observation, accurate mathematical calculation, unique seismological observation and data, advanced ship building technique, and gear wheel system, in particular her four major inventions--paper making, printing, compass and gunpowder, exerted a huge impact on the development of world science and technology and prompted the progress of the civilisation of the whole world. Her important scientific and technological achievements have won world recognition. The feats of her scientists are duly respected. Ancient Chinese science and technology have become the subject of study by the whole world. The ringlike mountains on the moon are named after Chinese scientists, Shi Shen, Zhang Heng, Zu Chongzhi, Guo Shoujing and Wan Hu.

The development of civilisation is a historic heritage, in which the best survives, carrying forward and gaining depth into knowledge as it progresses into the future. The history of science and technology is no exception. Having developed on a higher plane, deeper in depth and more extensive in scope, the achievements of science and technology today have grown on the basis of the ancients. It is difficult to thoroughly grasp the present if we do not know the past. Knowing the past helps us to forge ahead in greater strides into the future. In studying the history of ancient Chinese science and technology three aspects have to be stressed: we must unveil the inner laws on the relation between natural science and the development of engineering technology; study the social factors that promote or restrict activities of science and technology; and research the counteraction of science and technology on society with a view to promoting science and technology as the most vigorous productive force.

In 1997-98 the State Cultural Relics Bureau entrusted the National Museum of Chinese History to sponsor an exhibition on artefacts of ancient Chinese science and technology and compile this catalogue. Since so great and extensive are the achievements of ancient Chinese science and technology, it is not possible to exhaust them in a short account or in one book. Only a list of 10 subject matters, including four great Chinese inventions and astronomy, have been included. We shall nevertheless do our uttermost to do justice to the great achievements of our ancestors. The purpose is to learn from the old to better grasp the new. Our thanks are due to various provincial and city museums and cultural relics units for providing us with necessary and excellent artefacts for use by the exhibit.

Ma Zishu, Deputy Director,
State Cultural Relics Bureau 马自树

Aug. 1997

A GENERAL ACCOUNT OF THE DEVELOPMENT OF ANCIENT CHINESE SCIENCE AND TECHNOLOGY

To the primitive man, nature, a profound mystery, is difficult to predict or fathom. It is utterly impossible for him to live apart from nature—not even for one moment. He must study his environment (including the sky which is far removed), the natural substances and synthetic objects—to master, adapt, improve and invent, for the sake of existence and production. Natural science and engineering technology originated in primitive times, expanding and developing later. They are a summing up of the experience of man in the struggle to know and transform nature.

All nationalities and states on the globe have developed excellent science and culture, in spite of the fact that the culture of various regions may be deep or shallow in depth, broad or narrow in scope, and numerous or few in projects and the impact on the progress of mankind may be tremendous or minimal. They constitute the common wealth of mankind, an inalienable part of the history of global science and technology. They are truly the thousand streams and gullies that go to form a river, which empties itself into the sea. As one of four great countries with ancient civilisation, China is a big and magnificent river that has a long history and conglomerate of humanities or cultures.

As the biggest centre in which world agriculture originated, China grew many crops in the old days as she does now. She is the first country to develop industrial crops. The first systematic theoretical work on cultivation and farming in China is entitled *Qi Ming Yao Shu* (Essential Skills of the Common People), the earliest book on agriculture in the world now extant. As early as the final phase of primitive society ancient astronomy was rooted in China. Astronomical observation instruments of various sorts were invented in China, which first formulated the earliest stellar, most advanced calendar (at the time), astronomical records on the earliest eclipse of the sun, sunspot, the Halley Comet and Supernova.

Apart from these China is a centre for the development of mathematics in early days. Many theorems or solutions were first worked out in China: such as, Pythagorean theorem, 8 digital value of pi, equation of higher degree and numerical solution of simultaneous equation of higher degree, higher power binomial expansion of coefficient rule, summation of arithmetic series of higher degree, interpolation, linear congruence, and solution of equation of higher degree in several elements. Three thousand years ago China was the first country to have adopted the decimal system. *Mo Jing* (The Book of Mocius) of the Warring States period mentioned the definitions of circle, line and plane, which were similar to the definition of Euclidean geometry—earlier than the West by many years. Chinese achievements in physics were outstanding. The same book records the concept of inverted image by light through a small aperture, focus (this is something like the focus on a modern spherical mirror) and the centre of sphere. In other fields China scored very high achievements in lever principle, transmission and resonance of sound wave, hydromechanics, and the use of gear wheel system.

In chemistry China made wonderful gains in alchemy—making pills in a refining process. The book *Can Tong Qi* (This is a combination of *The Book of Changes, Huang Lao* and *Luo Huo*) by Wei Boyang of the Eastern Han Dynasty is the earliest work on the subject now extant in the world. In it he recorded the method of making synthetic valence for the first time in history. He noted the difference in the ratio of various substances in chemical changes. A remarkable achievement in Chinese alchemy is the successful manufacture of black gunpowder.

Traditional Chinese medicine and pharmacology have developed a school of their own, forming into an independent comprehensive system apart from the West. Books on medicine and pharmacology appeared in the country down through the ages, compiled by government or private scholars. Chinese books on medicinal herbs exceed by far books on botany by Western scholars in extent of knowledge and comprehensiveness. Books on acupuncture dealing with the theory of main and collateral channels (network of passages through which vital energy circulates and along which the acupuncture points are distributed) are outstanding contributions that shine in great splendour in the world.

Ancient Chinese architecture and buildings are known for their scientific and rational composition, magnificence and intricate skills. Chinese architecture includes city planning, palace, temple, pavilion, residential dwelling, pagoda, garden and park. Over 4,000 years ago the Xia Dynasty built halls and city walls of considerable size. The Qin Dynasty built the famous Great Wall, referred to in Chinese as the wall of 10,000 li. It was repaired, rebuilt or expanded by succeeding dynasties. Today the Great Wall stands in lofty style. The Tang Dynasty had the biggest metropolis of its time—Chang'an (now Xi'an), famous for its comprehensive layout. The majestic palaces of the Ming and Qing dynasties in Beijing win great admiration from the entire world for their superb workmanship, excelling nature. Chinese architecture is of great beauty and many buildings have survived to this day despite the ravages of time. *Ying Zao Fa Shi* (Building Formulas) by Li Jie of the Northern Song Dynasty, an important summing up of ancient Chinese architectural technology, is the most complete work on architecture in the world.

Science and technology complement each other, fulfilling separate but complementary needs. China led the world in sericulture and silk embroidery in primitive times. She retained her continuous lead in ship building, bronze smelting and casting, and wine making. Chinese forebears dug wells and irrigated cropland. The Du Jiang Yan water conservancy project was completed and became famous. Mining and smelting of iron and steel saw uninterrupted development. The magnetic compass, paper manufacture, printing and gunpowder, ceramics, and tea planting originated in China, which made the first ever seismograph to detect and record the movement and vibration of the earth.

What we have enumerated above encompass a very wide range of scientific and technological subjects--the fruit of labour and wisdom of Chinese labourers by the thousand, who made contributions to developing science and technology with themselves remaining in utter obscurity. Their life and work are nowhere to be found in books. Yet what they did constitute the solid foundation of ancient Chinese science and technology. Among them have emerged very eminent scientists. The rollcall of honour includes many: Mocius, Lu Ban, Bian Que of the Spring and Autumn and Warring States Period; Li Bing and his son, Zhang Heng, Cai Lun, Zhang Zhongjing and Hua Tuo of the Qin and Han dynasties; Ma Jun, Zu Chongzhi, Liu Hui, Ge Hong and Jia Sixie of the Wei, Jin and Southern and Northern Dynasties; Li Chun, Sun Simiao and Yi Xing of the Sui and Tang dynasties; Bi Sheng, Shen Kuo, Su Song, Tang Shenwei, Qin Jiushao, Yang Hui, Jia Xian, Li Ye, Zhu Shijie, Guo Shoujin and Wang Zhen of the Song and Yuan Dynasties; and Li Shizhen, Song Yingxing and Xu Guangqi of the Ming Dynasty.

Compared with four ancient countries—Babylon, Egypt, India and Greece, ancient China has special features in her scientific and technological achievements. She forms a system of her own. In ancient Chinese medicine, pharmacology and architecture, China differs markedly from the West in propounding theory, selecting materials, the source of medicine, technological process and method in healing sickness or disease. In astronomy, Chinese theories on cosmos are found in such works as *Gai Tian huo* (theory of canopy heavens), *Hun Tian Shuo* (theory of sphere heavens) and *Xuan Ye Shuo* (theory of expounding appearance in the night sky). The cosmos is described as containing three heavenly areas and 28 su or locations of stars. The tropics is used as the origin to co-ordinate the entire cosmos into 360 and one quarter degrees. Based on astronomical observations a lunar calendar has been worked out incorporating many advanced features as are rarely seen in the world.

Chinese mathematicians applied mathematics to practical work to measure the area of cropland whose unit was mu and to measure the volume of cellars, warehouses, gullies or dikes. Movable beads that can slide on counting frame appeared in the Western Han Dynasty. They developed into the abacus in the Song and Yuan dynasties. Up to now abacus is still used as a convenient calculating device with high efficiency. The situation in other science and technology is much the same. This is due to China's unique geographical location and environment and factors in economy, politics, ideology and culture, vastly different from those of Western countries. Major inventions, such as seismography, compass and chariot carrying drums that precede a royal cortege are necessitated by political needs. Medical and pharmacology books, calendar and works on agriculture, compiled by governments of successive dynasties, prompted the advancement of science and technology. Conversely due to political and military needs bans were imposed on the size of ships by limiting the number of sails and on ships leaving port for foreign countries. This led to a decline of ship building and navigation during the mid-Ming Dynasty. These are cases in which politics either promote or retard the development of science and technology.

The history of ancient Chinese science and technology has won the focus of attention of various countries for two reasons: One. Ancient Chinese science and technology have made great achievements. The English scholar Robert Temple in his book *China—Nation of Discovery and Invention,* enumerated 100 Chinese inventions (half of the world's basic inventions). This had great significance and exerted huge impact on the progress of human society. Two. The special pattern of the development of ancient Chinese science and technology possesses distinctive Oriental system, greatly enriching the contents of the history of global science and technology. Without a deep understanding of the history of ancient Chinese science and technology it would be difficult to understand the history of the science and technology of the world. Without an understanding of the former our understanding of the latter will be incomplete, to say the least.

Modern science and technology have forged ahead in great strides toan extent inconceivable to the ancients. As we trace the origin of enormous progress of the moderns we see that they have taken roots in the ancients, who provided valuable lessons for the moderns to learn. It is our hope that by exploring the laws of the development of ancient Chinese science and technology, understanding the social factors that either promote or hold back the progress of science and technology, and studying the impacts of science and technology on the progress of society, we shall be able to better grasp the new by knowing the old. We shall develop science and technology more deeply, on a higher plane, and at more extensive level in order to make science and technology the most invigorating productive force. We have chosen in this album ten subject matters as a brief introduction to the wealth of ancient Chinese science and technology for the benefit of our readers.

Seismograph (replica)
This is the first instrument in the world invented by the Chinese—Zhang Heng, Eastern Han Dynasty (132A.D.) to measure earthquake. Made of copper with a diameter of 8 chi (about 1.90 meters). it is shaped like a wine jar, whose centre is a stud that responds to shocks. There are eight sections of lever machinery. When there is earthquake one section will send a wave of vibration of earth movement that makes the dragon spit a pill into the toad below. The sound of pill dropping and the location in which pill drops tell earthquake and its direction. Highly sensitive, this seismograph can detect earthquake or vibration of earth movement not detected by man's own perception. The replica is made by Wang Zhenduo based on "Biography of Zhang Heng" in the *History of the Later Han Dynasty.*

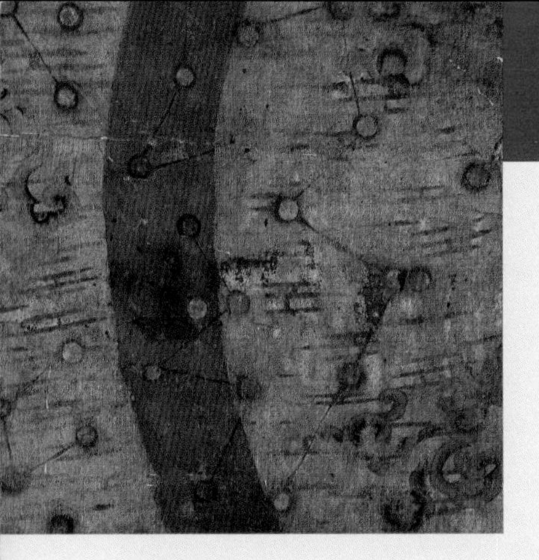

Ancient Chinese Science and Technology

ASTRONOMY

Astronomy has a long history in China, which has lots of books on culture recording her great achievements in astronomy. It keeps many artefacts of ancient astronomy, graphically reflecting the actual development of ancient Chinese astronomy.

As early as the New Stone Age Chinese forebears began their observation of heavenly bodies, inscribing pictographs of sun and moon on their pottery. Chinese observation of astronomical phenomena became more accurate and detailed during the Xia, Shang and Zhou dynasties. The eclipse of the sun was recorded 4,100 years ago—the first ever record in the world. The inscriptions on bones and tortoise shells of the Shang Dynasty had numerous records of the eclipse of the sun and moon plus the nova star in the Scorpio. Apart from these, there are words on inscriptions that give the meaning of sunshine, cloud, thunder, electricity, frost and snow, which bear testimony to Chinese concern with meteorology. Over years of long practice in life and production, the Shang people accumulated rich knowledge of astronomy and calendar. A year was divided into 12 months, with 13 months for the leap year. Based on the calculation of ten heavenly stems to record the movement of the sun made by the people of the Xia Dynasty, the earthly branches were added. The Western Zhou Dynasty calculated the leap year at year end, making a total of 13 months. The records of bronze inscriptions divide the month into four parts, known as chujie, ji shengba, jiwan and jisiba. At the end of each part are celestial stems and terrestrial branches. The inscriptions carry words, such as: dan, zhao and xi to indicate morning, day and evening. The sun may cast long or short shadows. This is measured by the tu gui, a clay measurement to ascertain the winter and summer solstices or the longest or shortest day in the year. The clepsydra is used to measure the time of day by trickling of water. It originated in late primitive society. Markings were added in the Shang Dynasty to the water clock to indicate hours. Observation of heavenly bodies became more accurate as time went by.The comet is a strange phenomenon and its appearance caused amazement and association among men. In 613 B.C. Halley comet was recorded for the first time by the Chinese. This record is acknowledged by the world as the first ever discovery. In the 4th century B.C. the *Gan Shi Xing Jing,* a book by Chinese astronomers, described the 120 fixed stars, being the first constellatory table in the world. The fixed stars are given 28 locations, a unique method of division. The lid of a lacquer casket in the No. 1 grave in Lei Gu Dun, Sui County, Hubei Province carries paintings of 28 su or star locations--providing evidence to the fact that the division had been used by the Chinese in the 5th century B.C. (end of the Spring and Autumn to the beginning of the Warring States Period). The Warring States calendar is known as the 4-division calendar. A year is calculated as having 365.25 days. This figure was used in the Julian calendar of the Romans, 300 years later.

In the 70 years between 246B.C.-177B.C. (Qin and Han dynasties), greater progress was made in astronomy in China, as testified by writings on silk, in which the position of five planets--Jupiter, Venus, Mercury, Mars and Saturn had been recorded. The silk piece, excavated from the Mawangdui grave of the Prince of Changsha, has pictures of positions of comets in various shapes. In the No. 8 grave in the ruins of Niya, Xinjiang a protective shoulder covering (of the Han-Jin period) was discovered, which had this to say in its inscriptions: "The five planets are discovered in the East, which brings benefit to China." In *Shi Ji* or *Records of the Historian,* the chapter on astronomy, titled "Tian Guan Shu," divides the cosmos into five areas, fitting in with the description in the silk fabric in Niya. As early as 28B.C. China detected sunspots or dark areas on the sun. It was recorded in the *Han Shu* or *History of the Han Dynasty*. The recorded date was earlier than that of the West by many years. The book recorded the discovery of a new star in the location known as fangsu in 134B.C. The new star (nova) is also recorded by the West. Chinese astronomical instruments were improved, with a new armillary sphere made by Zhang Heng, complete with round rings with markings of degrees of the complete cosmos plus a telescope. The equator is used as co-ordination. The armillary sphere was later used by the rest of the world as the basic co-ordination system. The celestial globe moves round once a day, syncronising with the movement of heavenly bodies. The sphere is full of stars. In daytime the movement of stars can be seen. The length of the day is made more accurate with the sundial—the artefact unearthed in Tuoketuo County, Inner Mongolia towards the end of the Qing Dynasty. The sundial improves the water draining clock. Several clocks were excavated in Mancheng, Hebei Province, in Xingping, Shaanxi Province, and in Yikezhao League of Hangai Banner, Inner Mongolia. The clock is of the single phase type. In early Han Dynasty a two-phase clock came into use. There were three-phase and four-phase clocks in the Jin and Tang dynasties respectively, according to literature.

The astronomical observatory in ancient China was known as qingtai. Called shentai in the Shang Dynasty, it was changed to lingtai in the Zhou Dynasty. The ruins of the eastern lingtai of the Eastern Han Dynasty stands in Yanshi, Henan Province today, being the largest ancient observatory in the world. Zhang Heng, the famous astronomer, was said to have worked here during his lifetime.

These rather advanced instruments provide excellent condition for ancient Chinese astronomers to observe heavenly bodies, measure, and detect the relation between the operation of the sun, moon and earth. Yu Xi (Western Jin) discovered precession of the equinoxes and determined winter solstice moves westward one degree every 50 years in the zodiac. In 462A.D. Zu Chongzhi (Southern Dynasties) counted 365.2428 days as one year, which was incorporated into the Ming Dynasty calendar making it all the more accurate. In the 6th century A.D. Zu Hengzhi of the Southern Dynasties discovered that polar star has a deviation of over one degree with the North Pole. Zhang Zixin of the Northern Dynasties discovered that the sun and planets are not equal or balanced in velocity of visual movement. Fast and slow changes occur in cycle. These led to major improvements in the calendar of the Sui and Tang dynasties. In early 8th century Yi Xing of the Tang Dynasty made a hydro generated astronomical clock and re-determined the position of the planets. He made many preparations to revise the calendar. He selected spots to make actual measurement of the length of one degree in the meridian. This was the first time in the world that such actual measurement was ever carried out. The Tang Dynasty made progress in the observation of constellation. A mural painting in the Dunhuang Grottoes reveals 1,350 stars. The drawing was made by means of projection with a round tube. Twelve sections were made for the stars around the equator, measured in accordance with the position of the sun in each month. The method is similar to modern projection. The picture of constellation of Qian Yuanguan, king of the state of Wuyue in the Five Dynasties, is a much treasured relic. Few words are inscribed on the painting but the position of stars is found to be far more accurate than previous drawings—certainly a valuable piece of artefact of ancient science.

Astronomical research reached its zenith in the Song-Yuan period. The constellation picture unearthed in the grave of Zhang Shiqing of the Liao Dynasty in Xuanhua, Hebei Province illustrates the traditional 28 star areas plus 12 palaces of ancient Babylon—a merger of Chinese and Western astronomy discovered for the first time in history. The Song Dynasty conducted five astronomical observations on massive scale. The fourth one, carried out in 1078-1085 (the reign of Song Shenzong Emperor), is contained in a constellation picture, inscribed on stone in 1247 preserved to this day. The picture has 1,434 stars. The borderlines between zodiac, moon's path and milky way are clearly demarcated. The constellations before the invention of the telescope in 1608 in the West contained at most 1,022 fixed stars. The West fell behind China in the number of stars discovered. The Song Dynasty discovered 4 supernovas. The most important is the one close to the Taurus discovered in 1054. The nova evolved into the modern Crab Nebula.

Great advance was made in astronomical instrument making in the Song-Yuan period. An armillary sphere was placed on top of a water generated framework by Su Song in 1088. A clock, astronomical observatory and markings of time were integrated into one unit, mounted on platform. Guo Shoujing made 13 instruments. His abridged armilla is a high precision instrument and easy to use—rather advanced instrument of the world at the time. He set 27 local observatories across China. Only two are left standing: one with building intact in Dengfeng County, Henan Province and the other near Jianguomen city gate, Beijing. The second one, rebuilt in 1442 has survived to this day. The original instruments are replaced by Qing-Dynasty instruments. An abridged armilla and the armillary sphere of the Ming Dynasty are kept in Zijinshan Observatory, Nanjing. New advances are made with the invention of time piece for precise time keeping, in addition to the water generated clock. It is a huge copper pot made in 1316 with four small ones for steadying water flow. Water drops from one pot to the next. A wooden indicator rises as the water accumulates, telling the time of day.

In the Northern Song Dynasty Shen Kuo proposed to adopt a calendar with the Beginning of Spring (corresponding to Feb. 3, 4 or 5 in the Gregorian calendar) as the first day of the first month in the lunar calendar. Each month has 30 or 31 days, alternating with the lunar month of 30 days. The 24 solar terms are integrated into 12 months. This fits in with the need of agriculture. The proposal of the merger of Gregorian and lunar calendars was, however, turned down for one reason or another. In the 20th century the English calendar by the name of Bernard Shaw used the same method. The Tong Tian calendar of 1199 used the tropical year as a unit of time equal to the period of one revolution of the earth around the sun measured between successive vernal equinoxes. It is reckoned as 365.2425 mean solar days, only short by 26 seconds. The Gregorian calendar, now popularly adopted by the world, uses the same data. It is, however, later than the Tong Tian calendar by nearly 400 years. The calendar compiled by Guo Shoujing uses the value of the length of tropical year with radial quadratrix as the main method to measure the movement of heavenly bodies, thereby raising the precision of the calendar calculation. Even more important is the fact that the contents of Chinese calendar include eclipse of the sun and moon, movement of planets, location of fixed stars, in addition to the arrangement of year, month and solar terms. In many respects the Chinese calendar resembles a modern astronomical calendar. In a word, ancient Chinese astronomy assumed a pre-eminent position in the world for many years. It is a splendid page in the history of the civilisation and development of the Chinese nation.

1-1

1-1 Inscriptions on ox bone in the reign of Wu Yi, Shang Dynasty record the eclipse of the sun. Measuring 17cm X 2.6cm, unearthed in Anyang, Henan Province, the bone is one of the earliest records in the world on the eclipse of the sun. The Chinese words: Kui You Zhen Ri Xi You Shi refer to the eclipse of the sun in the reigning years of Wu Yi based on actual observation. It is preserved in the National Museum of Chinese History in Beijing.

1-2 Rubbings from an ox bone inscription on eclipse of sun with annotations.

1-2

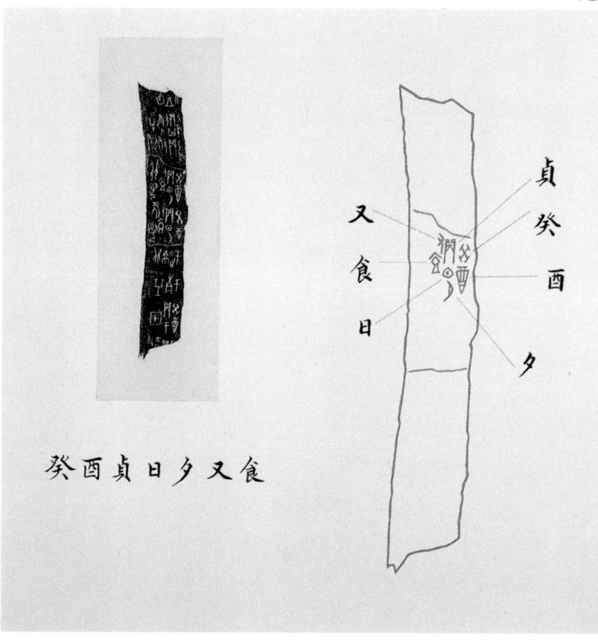

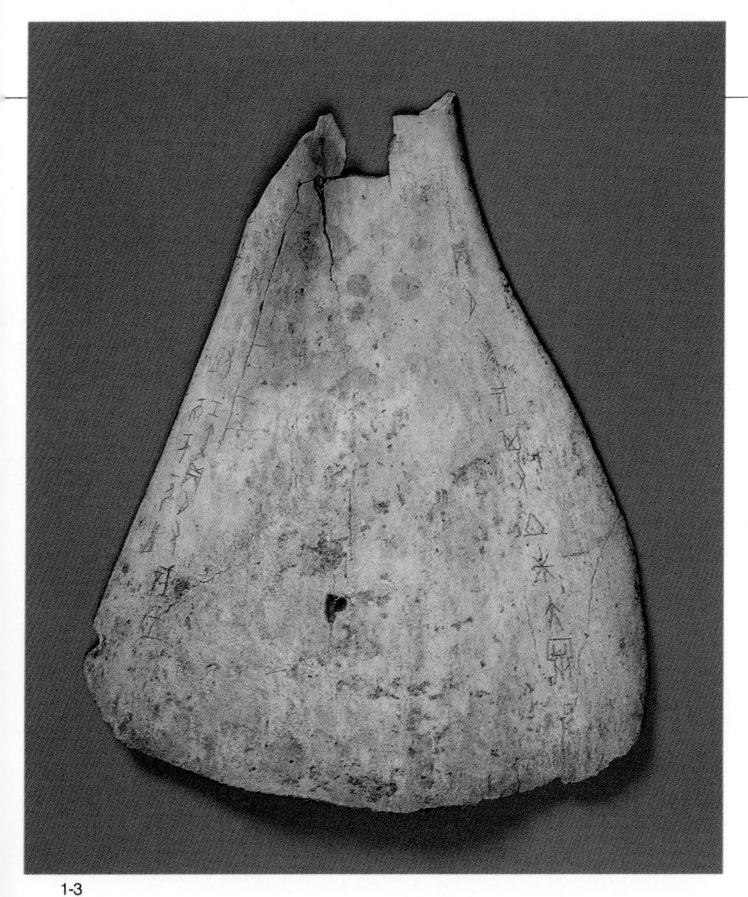

1-3

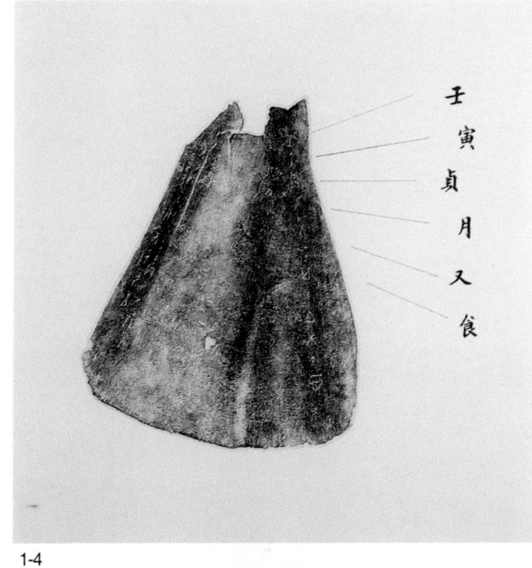

1-4

1-3 Yue You Shi ox bone unearthed in Xiaotun Nan Di, Anyang, Henan Province in 1973, the ox bone, 24.5cm high, 19.5cm wide at the bottom and 5cm at top, records the eclipse of the sun on July 2, 1173 in the reign of Emperor Wu Yi, Shang Dynasty. It is preserved in the Archaeology Institute, Academy of Social Sciences of China.

1-4 Rubbing from Yue You Shi inscription on ox bone showing eclipse of the moon.

1-5 Ox bone inscription, 22.5cmX6.8cm, on heavenly stems and earthly branches. It dates back to late Shang Dynasty (16th century B.C.-11th century B.C.). Modelled after Xia Dynasty calculation, Shang Dynasty calculations are based on a cycle of 60 years. The bone is inscribed with the heavenly stems and earthly branches, preserved in the National Museum of Chinese History. It was excavated from Anyang, Henan Province.

1-5

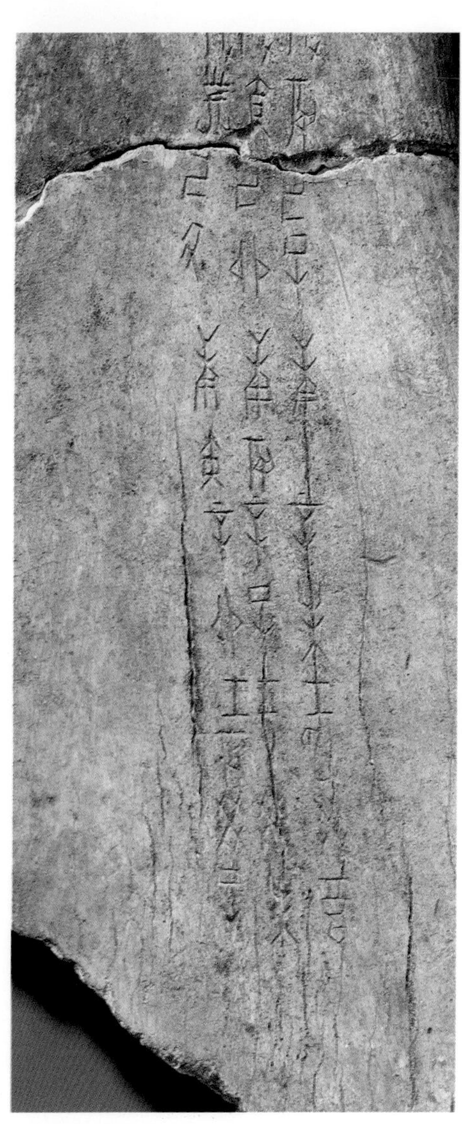

1-5*

1-6 Replica of tu gui or sundial (20cm high, each edge 35.5cm long and 2.3cm thick) of the Western Zhou Dynasty. Tu gui was used to measure the length of the shadow of the sun to ascertain winter and summer solstices and the four seasons of the year. It evolved into a stylus and plate for markings of the hour.

1-7 Part of painting of comets of the Western Han Dynasty, 150cm X 48cm.The original has 29 pictures of comets with accurate descriptions of 3 heads and 4 tails. It was unearthed from Mawangdui, Changsha in 1973 and is now kept in the Hunan Provincial Museum. There are 3 different heads and 4 tails, which testify to the accuracy of Chinese astronomical observation at the time.

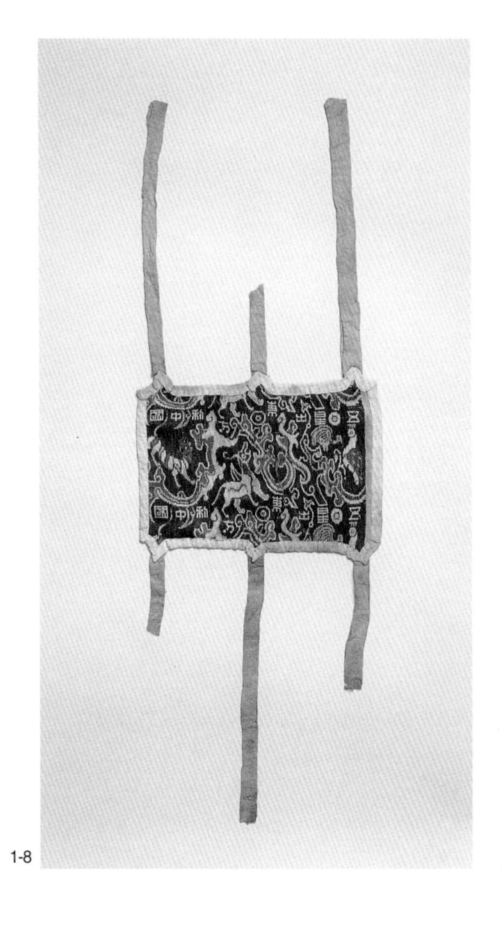

1-8

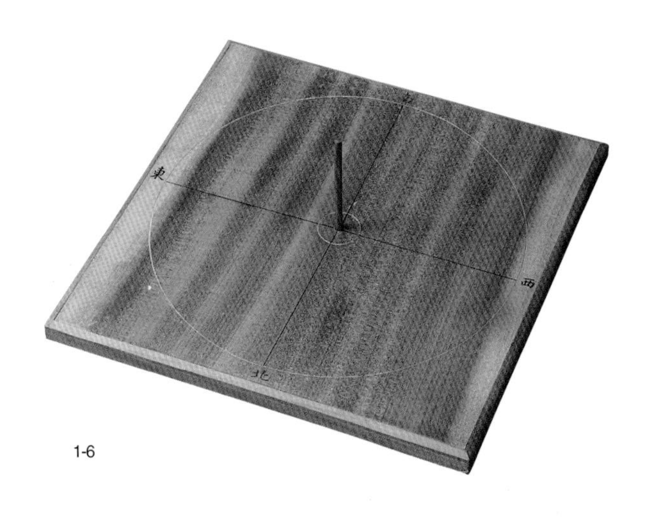

1-6

1-8 Silk shoulder covering, Han-Jin period, with picture of 5 planets detected in Orient, said to bring benefit to China. It is 18.5cmX12.5cm. The six silk bands measure 21cm. Unearthed from Niya ruins, Mingfeng County, Xinjiang Uygur Autonomous Region, it is kept in the Archaeology Institute of the Xinjiang Uygur Autonomous Region. The bands are made of coloured brocade and white silk cloth, with designs of peacock, crane, pi xie or spirit to ward off evils, etc. The style containing this theme is the only one unearthed so far in the world.

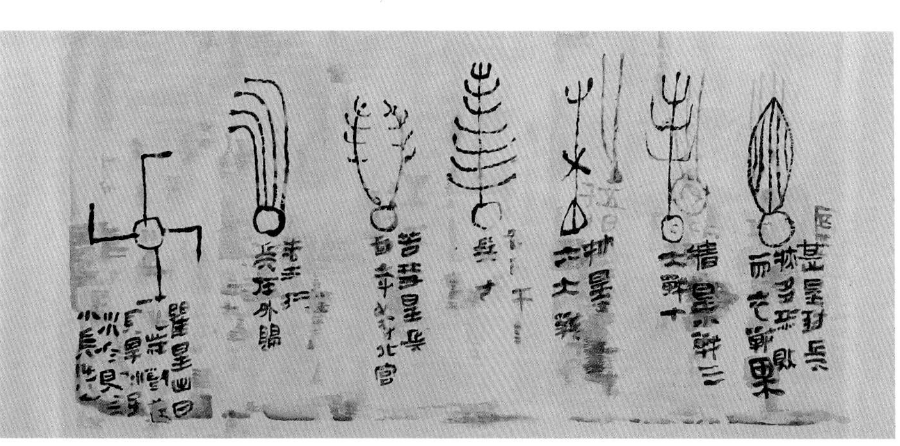

1-7

1-8*

1-9 Stone sundial 3.5cm high, and border of the square 27.6cm long (Han Dynasty). The plate is flat, smooth and square-shaped with a hole in middle. A circle with a radius of 4 cun is drawn plus 69 shallow holes, equidistant from each other. Outside are 69 numerals in lesser seal script. Excavated from Huhehot, Inner Mongolia in 1897, it is kept in the National Museum of Chinese History, Beijing.

1-10 Schematic diagram on sundial. The shadow moves from one numeral to the next, measuring length of day at sunrise, sunset and noon (and also length of night) in different seasons. TLV stand for NE, SE, NW and SW. T represents North, East, South and West. L is symbolic of carpenter square, a commonly seen decorative pattern of the Han Dynasty.

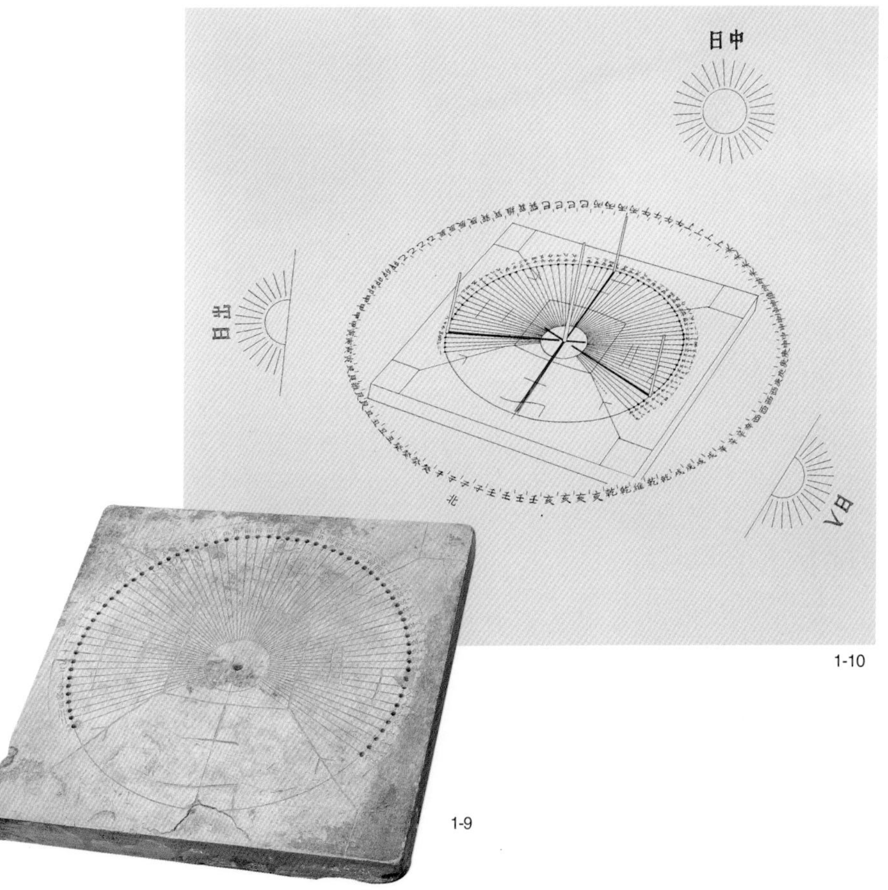

1-10

1-9

1-11*

1-11

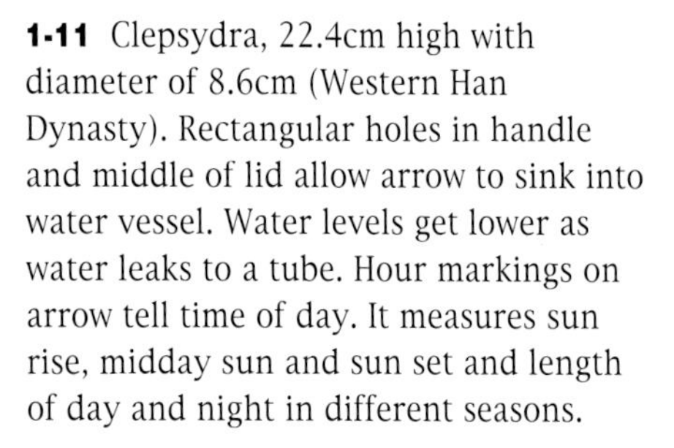

1-11 Clepsydra, 22.4cm high with diameter of 8.6cm (Western Han Dynasty). Rectangular holes in handle and middle of lid allow arrow to sink into water vessel. Water levels get lower as water leaks to a tube. Hour markings on arrow tell time of day. It measures sun rise, midday sun and sun set and length of day and night in different seasons.

1-12 Site of ancient observatory (staffed by 42 personnel) in Yanshi, Henan, where Zhang Heng used to work.

1-12

1-13 Rubbings of stone tablet showing astronomical chart with 1,434 stars. There are lines marking equator, zodiac, 28 star regions and milky way. Outer circle is 91.5cm in diamter. The star chart itself has a diameter of 85cm. The tablet is preserved in the Suzhou Stone Engraving Museum.

1-14 Constellation chart (in color) showing milky way in middle, flanked by clouds and 57 stars. The star regions are linked by threads.

1-15 Layout of observatory made by Su Song. In 3 tiers, it is 12m high with armillary sphere on upper deck. There are escape wheels in the water driven clock, first ever invention. The escapement principle is later used by modern clock.

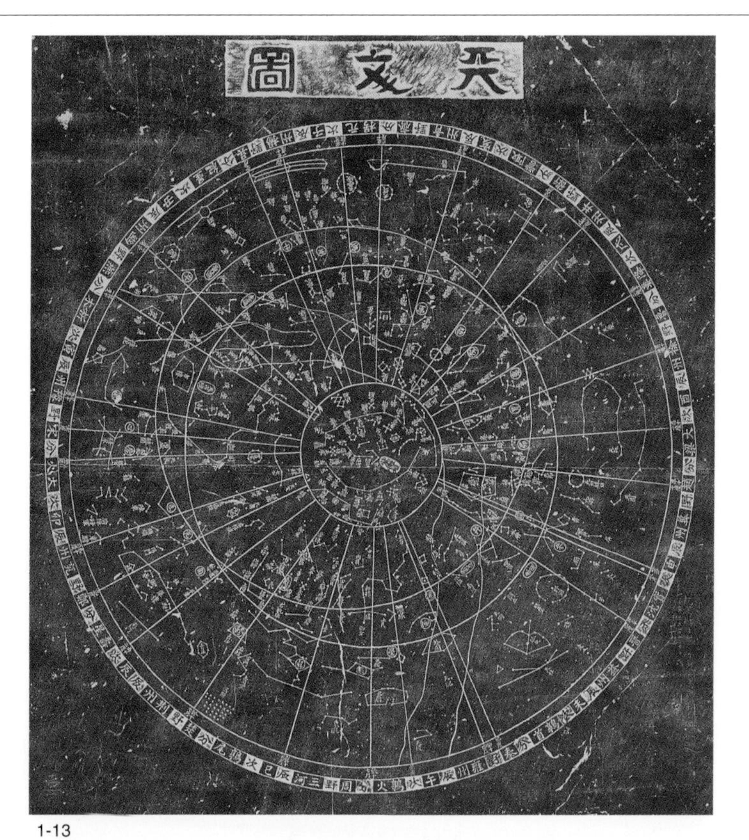

1-13

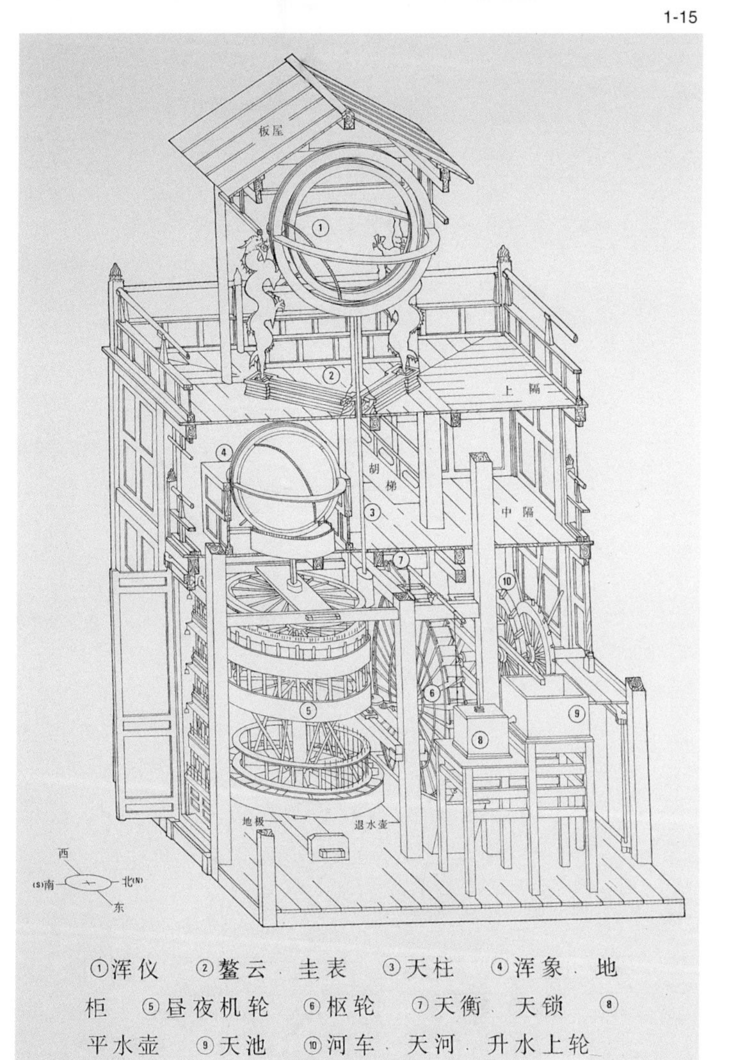
板屋
上隔
胡梯
中隔
退水壶
西
(S)南
北(N)
东
①浑仪 ②鳌云 圭表 ③天柱 ④浑象 地柜 ⑤昼夜机轮 ⑥枢轮 ⑦天衡 天锁 ⑧平水壶 ⑨天池 ⑩河车 天河 升水上轮

1-15

1-14

1-16

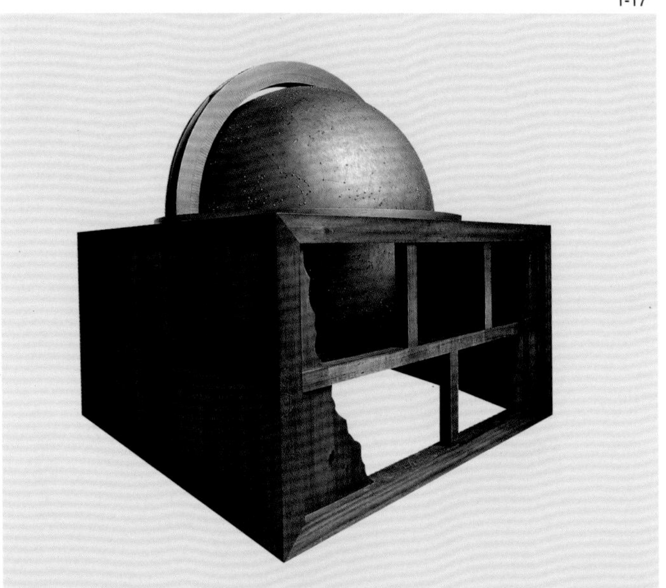

1-17

1-18

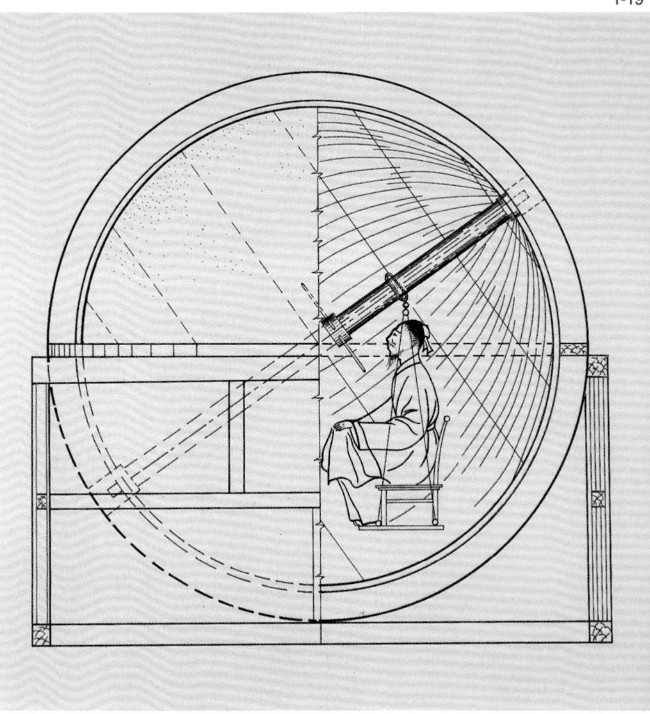

1-19

1-16 Replica of armillary sphere (water-driven) of Northern Song Dynasty. The original is made of copper. It is in three tiers. The outer tier is a six harmony instrument, measuring the length of day, etc. The middle tier measures four seasons and location of stars, sun and moon. The inner tier has a looking glass.

1-17 Armillary sphere with constellation. Latitude and longitude are seen in upper level. The lower part sinks into ground. The axle of the cabinet has an intersection angle of 35 degrees in relation to the ground.

1-18 This is a pseudo armillary sphere. A person sits in it and observes the universe. A handle enables him to turn the sphere round.

1-19 Layout of pseudo armillary sphere.

1-20

1-20 Replica of abridged armilla, 470 cmX328 cmX310 cm. The armilla is one of 13 instruments designed by Guo Shoujing around 1276. The armilla is an improved version and is convenient to observe the universe.

1-21 Replica of observatory in Dengfeng, Henan with a sundial. It measures shadow cast by sun to + (-) 0.2cm with 1/3 angle straggling, more precise than measurement of West made 300 years later.

1-21

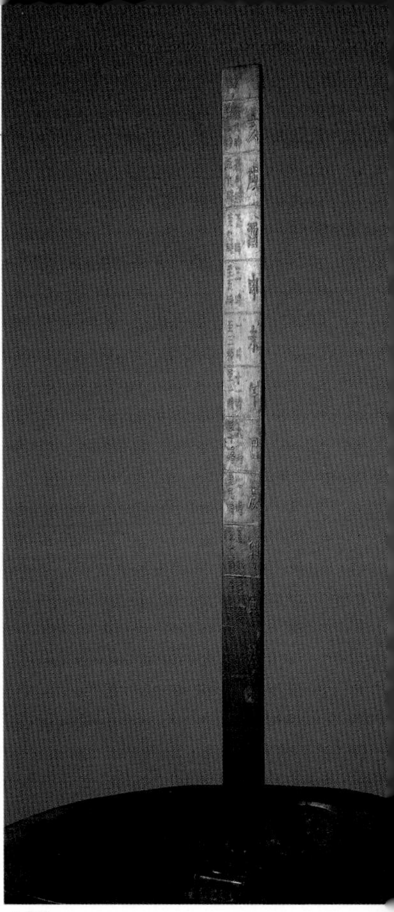

1-22*

1-22

1-22 Copper clock contains sun pot 75.9 cm high, moon pot 58.8 cm high, star pot 56 cm high and receptacle 76.3 cm. The copper clock was found originally on the city tower of Guangzhou. Water trickles down all the way to the receptacle. An arrow with markings registering time rises with the rise of the water level (preserved in the National Museum of Chinese History). Each pot, which has an overall height of 2.64 meters, has a lid. This clock is an early version of the instrument, which is divided into simple and complex types.

1-23 Observatory in Beijing is 14 m high, 20.4 m long in north -south direction, 23.9 m in east-west direction. There are 8 instruments including equatorial armillary sphere, heavenly body instrument, quadrant, ecliptic armillary sphere, horizon circle, zenith sector, sextant and elaborate equatorial armillary sphere rebuilt in early Qing Dynasty. The observatory was built in 1442 and stood on top of the southeastern city wall of Beijing. It served as the national observatory of the Ming and later Qing Dynasties. The original instruments of the Ming Dynasty have been destroyed or removed elsewhere.

1-23

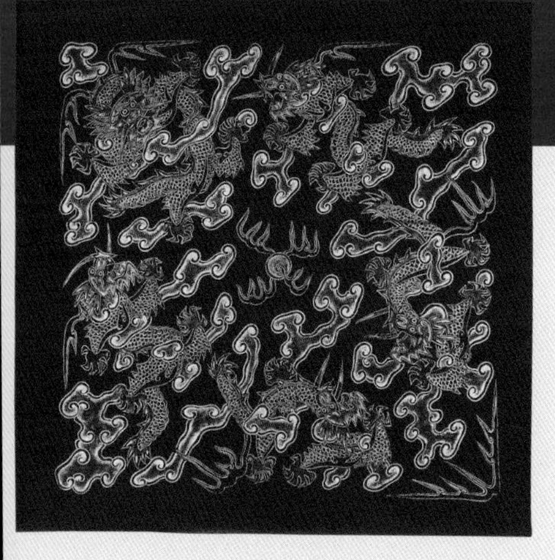

Ancient Chinese Science and Technology

PAPER MAKING

Paper making, one of China's four great inventions, brought fundamental change to writing , exerting a huge impact on the development of civilisation and culture of the world. A map drawn on paper was unearthed in the No. 5 Han Dynasty grave in Fangmatan, Tianshui, Gansu Province. Hemp papers of the Western Han Dynasty were excavated in Jingguan, Gansu and at the site of a beacon tower in Maquanwan, Dunhuang. Archaeology data show that paper making was invented in the Western Han Dynasty. Crucial changes occurred in paper manufacture in the Eastern Han Dynasty, when Cai Lun, Marquis of Longting (?-121A.D.), summed up past experiences of paper manufacture and improved technology, using bark of trees, broken bits of hemp threads, worn-out cloth and second-hand fishing nets. This improved the quality of paper and facilitated writing a great deal. Paper window screen technique was invented in the Wei, Jin and Southern-Northern Dynasties. The screen could be scrolled up and down, complete with frame and borders. Efficiency in paper production enhanced tremendously by the Eastern Jin Dynasty, as paper replaced silk as writing material. Glueing was invented in the Jin Dynasty, producing paper with adhesive quality. Paper made with glue in the Later Qin Dynasty (384 A. D.) is the earliest such paper now extant. A layer of starch was placed on the surface of the paper to make it smooth by polishing on fine stone to increase the strength of paper fibre and resistance to water. The glueing method was added to pulp making. Gelatin from plants or animals plus alum were used. The paper was known as shuzhi . According to an ancient Chinese literature dated 349A.D. unearthed in Xinjiang , tubuzhi (paper made with mineral powder from kaolin, talcum or gelatin) appeared in the Jin and Southern-Northern Dynasties. The gelatin was applied on the surface of the paper to improve its smoothness or whiteness. Paper made in color was called huang, which improved the quality still more. Paper was dyed with amur cork, which contained berberine that prevents it from decaying or from being eaten by moth. By Wei, Jin, and Southern-Northern Dynasties great stride was made in paper manufacture as thickness of pulp and scattering of fibre, density and smoothness were featured. Paper made from mulberry, creeper or bamboo was found. Paper production expanded as Xinjiang became a paper producing area at this time. Output was expanded all the more. Paper began to be used for a greater variety of purposes—writing, painting, window screen, lantern, cloth, cap, bedroll and armour. Wax treated yellow paper (tubuzhi treated with wax) was introduced in the Tang Dynasty. Adornment methods of lacquer and silk workers were used to sprinkle paper with gold or silver powder. This was known by a host of names: gold flower paper, silver flower paper or cold gold paper. Water wave paper and yahua paper with glossy surface came into use. Famous papers of the Tang Dynasty include yubanxuan paper, made of sandalwood bark , which was presented to the emperor as gift. It was also called xuan paper as it was manufactured in Xuanzhou. Anhui Province. The Southern Tang of the Five Dynasties produced chengxintang paper, which was white, smooth, refined and very tough.

Paper manufacturing technology reached maturity in Song and Yuan period as a new type of paper made of wheat stalk and rice stem began to be put into production. Used papers were recycled and used as pulp. Paper in large size—3 zhang or approximately 30 feet, was made possible due to the application of water generated hammer to make pulp. Plants(sunflower or carambola) were used as floating solution into the pulp.

Paper was used for a greater variety of purposes in the Song-Yuan period. Apart from writing and painting, even more papers were used to print books and paper currency. The development of firearms called for the manufacture of wrapping paper or cartridge bag for the gun. Paper used for writing letters called xie gong jian won renown in the Song Dynasty. Another famous brand was the paper for recording Buddhist sutras made by the Jin Su Temple in Haiyan, Zhejiang province. The famous brand paper of the Yuan Dynasty was the mingrendianzhi.

Books on paper manufacture appeared in the Song Dynasty. *Zhi Pu,* by Su Yijian, was published in 986A. D.—the first book published in the world devoted to paper manufacture. The Ming and Qing Dynasties saw comprehensive achievement in paper technology, when greater variety and better quality papers were put out as imitation of famous papers of preceding dynasties. New processed paper was made. Output of bamboo paper was highest in the Ming Dynasty. The famous Ming Dynasty Xuan De paper, ranking top as processed paper, was as famous as Xuan De boiler and Xuan De ceramics. Processed paper in the succeeding dynasty—the Qing, was most numerous, particularly in the reigns of Kang Xi and Qianlong, including plum letter writing paper, square in shape, adorned with powder wax plus designs in clay gold and silver pigment which was simply gorgeous. The list includes yaguang paper(with hidden flowery design), wood grain paper, fa jian, mica paper, wall paper and barrel paper. The *Jiangxi Da zhi: Chu Zhi* by Wang Zongmu and *Tian Gong Kai Wu: Sha Qing* by Song Yingxing are important works on paper technology during the Ming Dynasty. With the successful invention of paper manufacturing technique in China, the technology spread to the rest of the world as a result of economic and cultural exchanges. It first spread to Korea and Vietnam from China in the Jin and Southern-Northern Dynasties and later to Japan via Korea. In 751 A. D. a Chinese expeditionary army led by Tang Dynasty general, Gao Xianzhi, suffered reverses in battle in Central Asia. The Moslem army took captive a Chinese soldier, who was a paper manufacturing technician. He was sent to Samarkand to make paper for the Arabs. Meanwhile paper manufacturing technology spread to Central Asia. In the 1130s Arabs dominated West Asia and North Africa. Thanks to the Chinese the Arabs began paper making, which was spread to Europe, where it became popular in the 17th century. By the 19th century paper manufacture spread to all five continents in the world. China's paper making technology is certainly a great contribution to the culture of the world.

2-2

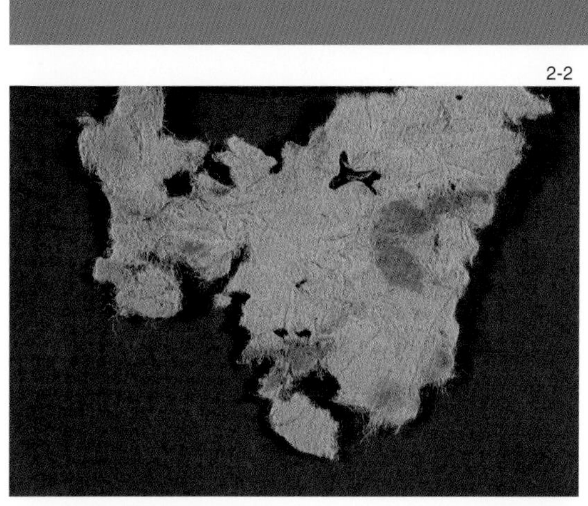

2-3

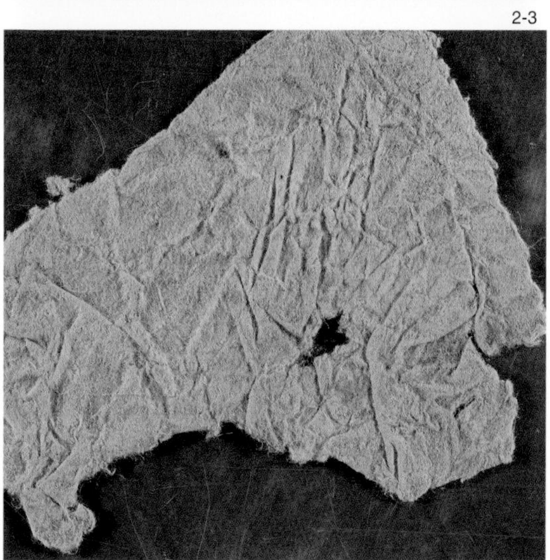

2-1 Western-Han map drawn on hemp paper, 8cm long, has a plane and glossy surface with fibre remains. Unearthed in Fangmatan, Tianshui, Gansu Province in 1986, it is the earliest paper map in the world, kept in the Gansu Provincial Museum. Ink lines indicate mountain, river, cliff and road.

2-2 Jingguan paper, 15cmX8.6cm, is white, thick and smooth, with tough texture. Excavated from Jinta County, Gansu in 1973 it is kept in the Gansu Provincial Museum.

2-3 Han-Dynasty Maquanwan paper, 14.5cmX10cm, was unearthed from the Maquanwan beacon tower site in Dunhuang, Gansu Province in 1979. The paper is of fine quality, white with evenly distributed hemp fibre.

2-4

2-4 Portrait of Cai Lun , who summed up previous paper making experience and introduced bark, hemp, used cloth and discarded fishing new as raw materials for paper making. This new kind of paper, named after him and with improved quality, replaced silk as material for writing in China.

2-5 What remains of a Buddhist sutra written on paper, 24cmX142cm. The quality of this Northern Liang paper is clearly improved. It is kept in the National Museum of Chinese History.

2-6 Tang Dynasty gelatine paper, 28.7cmX13.2cm, is water resistant and used for writing and painting. Excavated from Turpan, Xinjiang in 1928, it is kept in the National Museum of Chinese History.

2-7 Sutra on Tang Dynasty paper, 100cmX25.9cm with closely woven fibre texture. Unearthed from the former site of the Turpanharahezhuo city in Xinjiang in 1928, it is kept in the National Museum of Chinese History.

2-8 Page from Song Dynasty book, titled *Song Ren Cheng Shi Lun Ce,* 24.2cmX27.9cm. The Song Dynasty witnessed maturity in paper making skill. Paper was used to print books, in addition to drawing or painting.

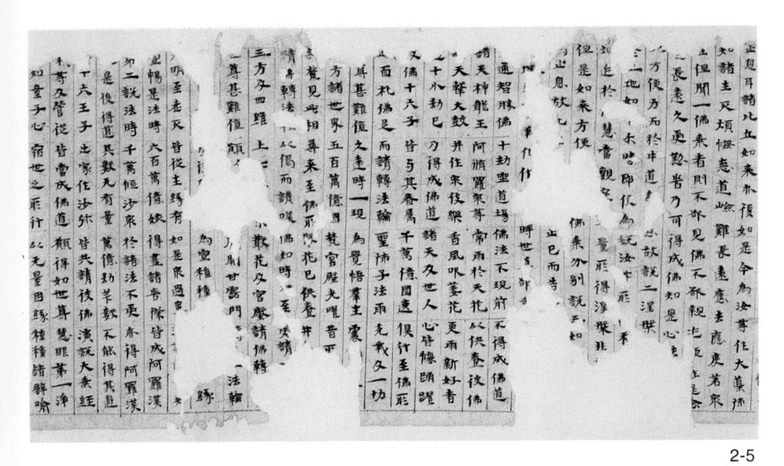

2-5

2-6

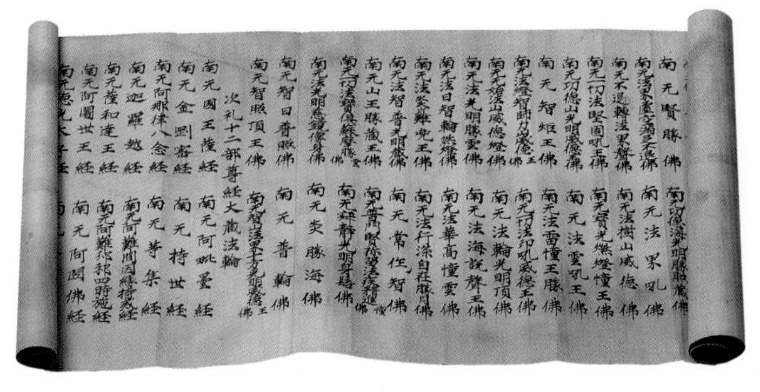

2-7

2-8

2-9

2-12

2-10

2-13

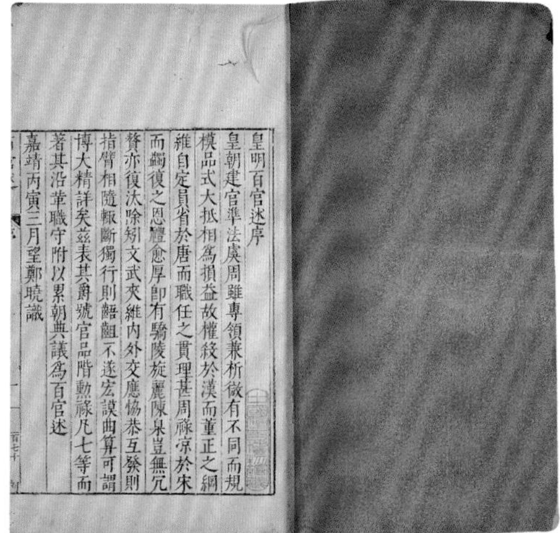

皇明百官述序

皇朝建官準法虞周雖專領兼析微有不同而規
模品式大抵相為損益故權紋於漢而董正之綱
維自定員省於唐而職任之貫理甚周祿京於宋
而蠲復之恩體愈厚即有驕陵旋麗陳臬豈無冗
贅亦復汰除剏文武夾維內外交應協恭互發則
指臂相隨輒斷獨行則齟齬不遂宏謨曲算可謂
博大精詳矣茲表其爵號官品階勳祿凡七等而
著其沿革職守附以累朝典議為百官述

嘉靖丙寅三月望鄭曉識

2-11

2-9 Ming Dynasty bamboo paper, 35.7cmX126.6cm. Bamboo was crushed, boiled and dried before it was used as pulp to make paper. By the Ming Dynasty paper was greatly improved in quality with the use of bamboo. Jiangxi, Fujian, Zhejiang, Anhui, Guangdong and Sichuan, which grow lots of bamboo, became prosperous bamboo paper producing areas.

2-10 Qing Dynasty xuan paper, 64cmX133.5cm made chiefly from bark of sandalwood mixed with carambola and straw. Xuan paper is a famous Chinese paper used for writing and painting.

2-11 Qing Dynasty moth-resistant flyleaf, 26.3cmX15.2cm. Many ancient Chinese books have this sort of flyleaf and are preserved till this day. This particular paper is kept in the National Museum of Chinese History. Known as "red for 10,000 years," the paper was attached to books to resist moth.

2-12 Qing Dynasty wax paper with ancient coinage design, 128.6cmX 68cm. It is kept in the National Museum of Chinese History.

2-13 Glossy mica paper, 88.5cmX62.5cm with coloured fibre texture. It is kept in the National Museum of Chinese History.

2-14 Square paper, 56.5cmX56.5cm, with cloud and dragon design. It is kept in the National Museum of Chinese History.

2-15 Schematic diagram showing Chinese paper technology spreading to the world by various routes. Chinese paper making skill first spread to Korea, Vietnam, Japan, India and Pakistan. By 8th-11th centuries it spread to West Asia and North Africa. After the 12th century it spread to as far as the American continent.

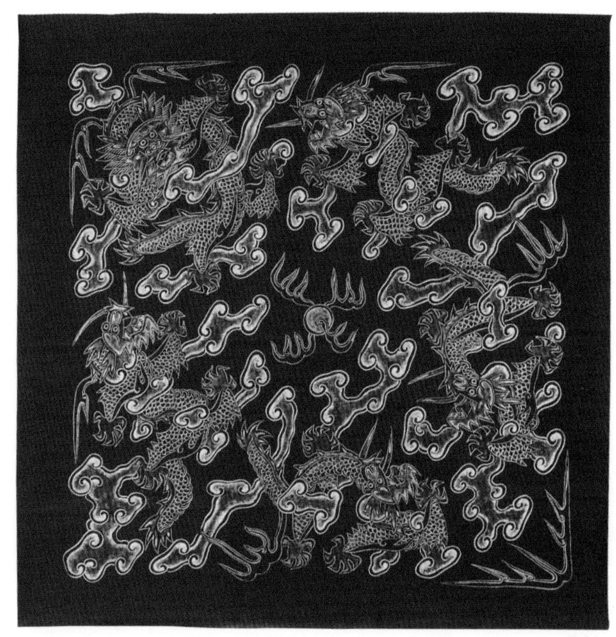

2-14

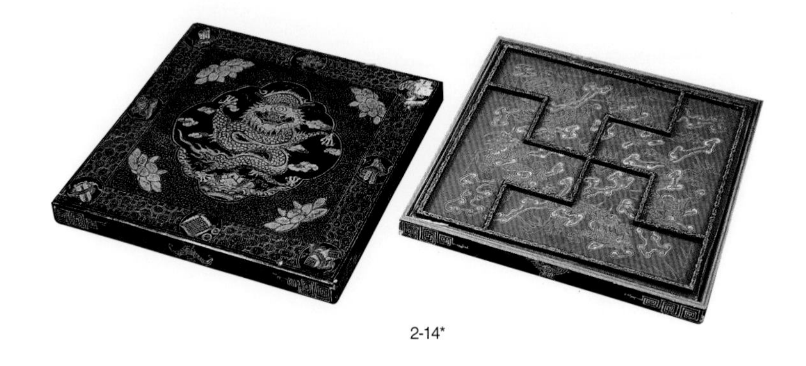

2-14*

2-15

Ancient Chinese Science and Technology

PRINTING

Printing, one of four great Chinese inventions, underwent two important stages of development: wood block cutting and letter press (movable type). The invention of printing owes its origin to the closely related seal, imprints of chops and printed marks, rubbings from stone tablets and printed designs on textiles. Carved stone tablets contained positive words in relief or raised type. Rubbings also contained positive characters. The earliest seal engraving is traced to the Western Zhou Dynasty. Stone inscriptions contained positive letters set in relief. Rubbings from inscriptions also had positive letters. The basic principle of the two is similar to printing. It may be said that seal engraving and tablet inscription are cutting blocks on small scale. The manufacturing method and process of engraving paved the way for printing, technologically speaking. The basic program in cutting block printing was to write down the words first and paste them on a wooden block (preferably wood from pear or date tree). Using knife the printer carved words in relief. This was the printing block.He then added ink and paper and pressed a brush onto the carved words. What he got from this constituted a printed matter.

The actual invention of cutting block was made in early Tang Dynasty. A Dharani sutra unearthed in the Sakyamuni Pagoda, Kyongju, South Korea, printed before 751A.D., is the earliest printed copy (with ascertainable date) now extant, spread from China to what was then the Kingdom of Xinluo in the Korean Peninsula. Slightly later editions of the Dharani sutra unearthed in Xi'an, Shaanxi Province of China during excavation in the Xi'an Metallurgical Machinery Plant and the Diamond sutra in Dunhuang as well as another Diamond sutra unearthed in Chengdu, Sichuan Province (dating back to 868) show fine engraving or cutting, which reached maturity in technology.

In the Song-Yuan period printing technology forged ahead, attaining a higher level of development. Books printed at this time were great in number with fine quality of printing. The words were written and cut in block letters. Such famous works of Chinese literature as *Tai Ping Yu Lan* (Taiping Imperial Encyclopaedia), *Tai Ping Guang ji* (Taiping Miscellany), *Wen Yuan Ying Hua* (Choice Blossoms from the Garden of Literature), *Ce Fu Yuan Gui* (Historical Anecdotes of Various Dynasties) and *Zi Chi Tong Jian* (History As a Mirror) were all engraved at government printing workshops. Private engravers and printers were set up at home. Private printing workshops also appeared. There were five engraving and printing centres in China: Henan (then known as Bianliang), Zhejiang, Sichuan, Fujian and Jiangxi. New development was made in cut block printing in 1264-1294 (Yuan Dynasty), with the emergence of chromatography in red and black. This was a simple process at first. The colours were pasted on different areas of the block for printing. By the Ming Dynasty a wider range of colour came into use on separate blocks. This traditional method was used for a long time until the Ming and Qing dynasties.

Wood-cutting art reflects another aspect of printing. In the Ming and Qing period wood-cutting made splendid achievement. Six different colours emerged alongside editions in showy and flashy expressions in writing and curved designs slightly raised in middle like an arch. The technique is superb. An imitation made skilfully in printing makes one believe that it is the original or genuine object. The representative works are the Luo Xuan Bian Gu Jian Pu (Ming Dynasty), a watercolour chromatography, Shi Zhu Zhi Hua Pu or an album of paintings by Hu Zhengyan of Huizhou, Anhui Province (1644) and another album of paintings known as Jie Zi Yuan Hua Pu(printed in the reign of Kangxi, Qing Dynasty).

Cut block printing technology played a crucial role in spreading culture. One page of print required a block of wood, which was expensive. Woods are not easy to preserve. Improvements had to be made as this process spread in the country to answer the need of society. The movable type in printing appeared.

Movable type had its beginning as far back as the Spring and Autumn Period. The bronze inscription on a circular vessel called gui (used as container for grain) of the Duke of Qin involved the use of letter press, cast into bronze mould. Since no paper had been invented plus a host of other reasons, it never developed into printing industry. In 1041-1048 Bi Sheng invented the movable type, using adhesive clay as single character mould. The characters were engraved in reverse and fired into pottery to make movable type. Two plates were used alternately during type setting and printing. The characters were taken out one by one and put away to be used next time. Raising the efficiency of printing, the movable type was a revolution. Just exactly what had Bi Sheng printed with his method, no records can tell us. Nor are there any artefacts to tell the story. The earliest printed matter from clay letter press was the *Wei Mo Jie Suo Shuo Jing* (Sutra by Vimalakirti) unearthed in Heishui, Ningxia region of China. The copy found its way to the St. Petersburg city library, Russia. The sutra was written in Xi Xia and Han Chinese languages. It was printed by the Xi Xia Dynasty, not too far distant from the time of Bi Sheng. Bi Sheng used movable type made of wood. But the performance was none too successful. Works titled *De Xing Ji* (a book on moral integrity and deeds), *San Dai Xiang Zhao Yan Ji Wen* and *Bai Fa Ming Jing Ji* (One Hundred Ways in Clear Mirror)were printed by movable type. In 1991 a copy of a work titled *Ji Xiang Bian Zhi He Ben Xu Juan Di San* was unearthed in Bai Si Gou Square Pagoda, Ningxia. This work printed in Xi Xia Dynasty on wood is the earliest printed matter of the movable type in the world.

Wang Zhen of the Yuan Dynasty perfected the wooden movable type technique. Serving as magistrate of Jingde County, he ordered workmen to carve 30,000 words in Chinese in the course of two years. In 1298 he spent a little over one month to print the 60,000-word *History of Jingde County*. To raise the efficiency of type setting Wang Zhen made a revolving plate, on which letter press was placed. One man read the manuscript while another pushed the wooden revolving frame and took out the required letter. This reduced the workload and bolstered performance. Wang Zhen summed up his experiences in his *Zao Huo Zi Yin Shu Fa* (a book on ways of making movable type in printing), which was appended to his book *Nong Shu* (On Agriculture). It provides valuable information on the history of printing. Towards the end of the 13th century movable type spread far and wide across China. The best example is a book in Uygur language, printed in movable type preserved to this day. In early Yuan Dynasty tin was experimented to make movable type. Its performance did not come up to expectations. It is the earliest movable type in metal in the world. In 1490 (Ming Dynsty) a Mr. Hua of Wuxi, Jiangsu Province used copper as movable type and printed 50 copies of *Song Zhu Chen Zou Yi* (a book on petitions to the emperor by various officials in the Song Dynasty) in 150 volumes. Stereotype in lead appeared around this time. Deserving particular mention is the *Gu Jin Tu Shu Ji Cheng* (Collection of Ancient and Modern Books) in 10,000 volumes, packed into 5,020 books by the Imperial Household Department. It runs to 160 million words—the largest encyclopedia in the world at the time.

Printing played a crucial role in spreading culture to the world, facilitating cultural exchanges. From China it first spread to Korea, Japan, Vietnam and Ryukyu Islands and Southeast Asian countries. The great collection of Buddhist scriptures or the Tripitaka (*Da Zang Jing*) was printed in Korea in the 9th century. Japan first printed Dharani scripture in the 8th century. She put out *Wei Shi Lun* (On Cognition) in the 11th century. Printing spread westward to areas dominated by the Arabs. In the 13th century paper currency was printed in Persia. Both Arabian and Chinese languages appeared on Persian paper currency. Printing spread to Europe. At the turn of the 14th and 15th centuries paper cards were printed by means of movable type in the southern parts of Germany and Venice,Italy. The movable type was used by the German Johann Guttenberg, with ink on raised surface pressed on to paper. He printed the Holy Bible in 1456 which had 42 lines to the page.

3-1

3-1*

3-2

3-3

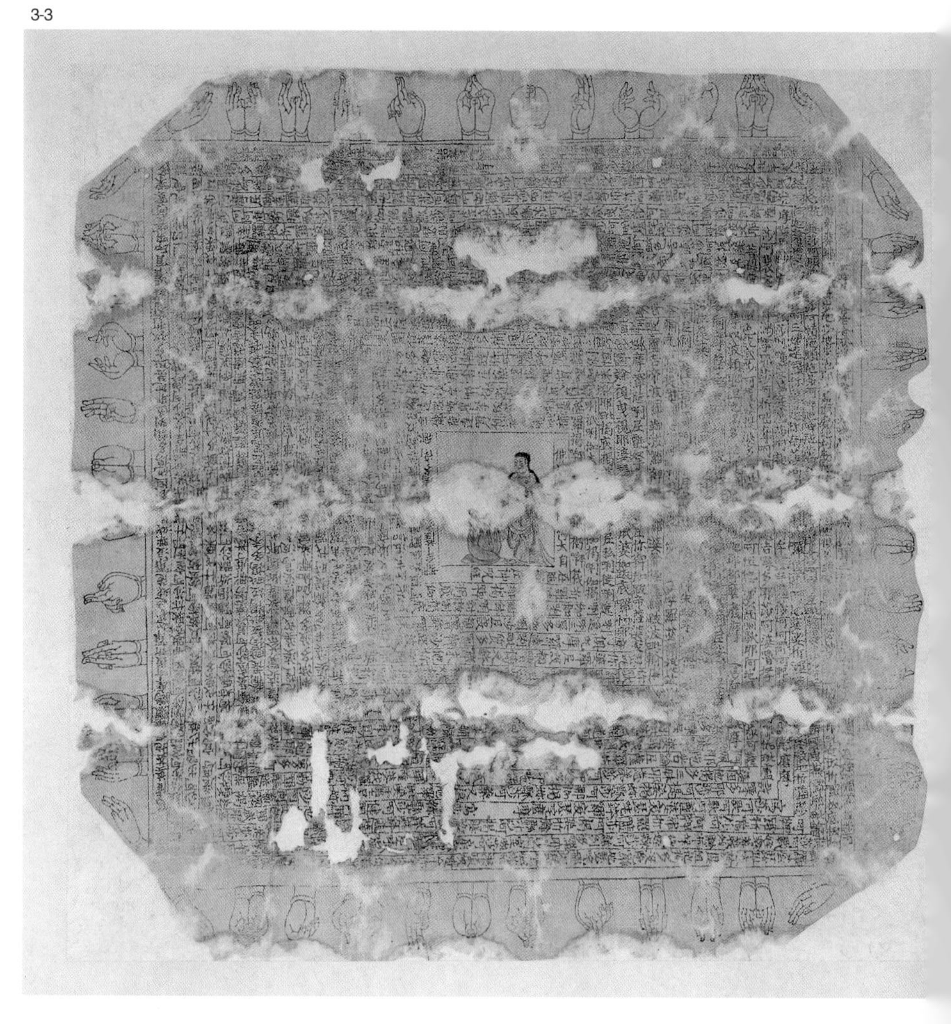

3-1 Lao Yang Si Kou-Warring States copper seal, 1. cubic cm. Seals can be traced to the Zhou Dynasty and was set in intaglio or relief, which applied the same principle as printing. This relic is preserved in the National Museum of Chinese History.

3-2 A Western Han Dynasty clay storage for keeping letters and documents, 1.1cm high with a diameter of 2.6cm. Excavated from Tengxian, Shandong Province, it is preserved in the Museum of Chinese History. Clay storage was popularly used in the Warring States Period and Qin and Han dynasties. They kept letters and documents and were tied together with a rope and clay. Words were printed on the clay or gumming dirt.

3-4

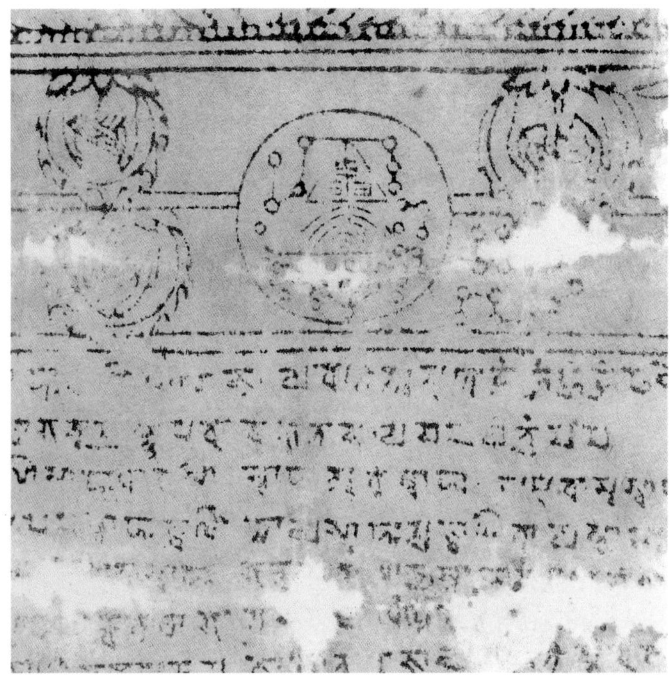

3-4*

3-3*

3-3 A page from Dharani scripture in Chinese belonging to mid and late Tang period, 35cmX35cm, unearthed in the construction site of the Xi'an Metallurgical Machinery Plant. It is now preserved in the Xi'an City Cultural Relics Management Bureau.

3-4 Tang Dynasty Dharani scripture in Sanskrit, 27cmX26cm, unearthed is 1974 in the construction site of the Xi'an Diesel Machinery Plant. It is preserved in the Xi'an City Cultural Relics Management Bureau.

凡欲讀經先念淨口業真言遍
脩唎 脩唎 摩訶脩唎 脩脩唎 娑婆訶
奉請除灾金剛 奉請辟毒金剛 奉請黄随求金剛
奉請白淨水金剛 奉請赤聲金剛 奉請定除厄金剛
奉請紫賢金剛 奉請大神金剛
金剛般若波羅蜜經
如是我聞一時佛在舍衛國祇樹給孤獨園與大
比丘衆千二百五十人俱尒時世尊食時著衣持

3-5

3-6

3-5 Tang Dynasty Diamond sutra discovered in Dunhuang Grottoes. Printed in 868 it is in scroll edition. There is an illustration on the preaching of the sutra. The sutra is at the back of the illustration. This cultural relic found its way to the West and is now kept in the British Museum, London.

3-6 Tang Dynasty Dharani scripture, 30.5cmX34 cm, printed on mulberry paper, semi-transparent. Unearthed from Wangjianglou, Chengdu, Sichuan Province in 1944, it is kept in the National Museum of Chinese History.

3-7

3-7 Zhi Sheng Guang Jiu Yao Tu, copy of Liao Dynasty earliest wood engraved block using ink on bark paper. Unearthed from the Light of Buddha Temple, Yingxian, Shanxi, it is the earliest, largest and best Chinese printed matter in colour from wood engraving. It measures 94.6cmX 50cm.

3-8

3-9

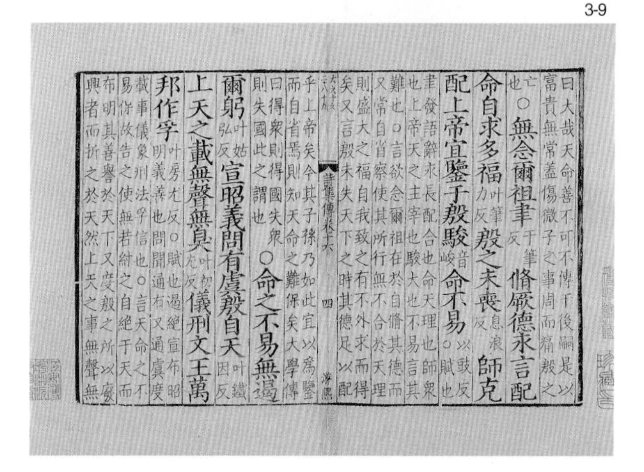

3-10

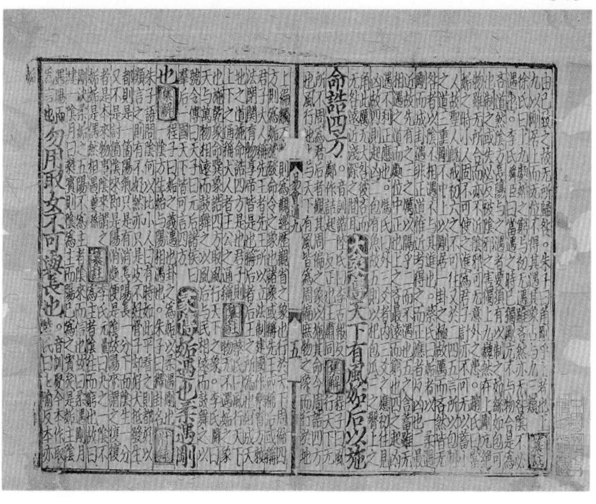

3-8 Jinan Liu Jia Gong Fu Zhen Pu (needling skills of the Liu family in Jinan), 12.4cmX13.2cm, trade mark in copper mould, Song Dynasty. It is preserved in the National Museum of Chinese History.

3-9 Page from Song Dynasty book: *Shi Ji Zhuan* (On Poems in 20 volumes) compiled by Zhu Xi, looking like a fresh print. Measuring 31.8cmX44.6cm, it is kept in the National Museum of Chinese History. It is an important work and a rare one on *The Book of Poetry* by Song scholars.

3-10 *Zhou Yi Hui Tong* (On"The Book of Changes"), 23.8cmX29.6cm, Yuan Dynasty relic. It is preserved in the National Museum of Chinese History.

3-11

3-12*

3-11 Paper currency of the Ming Dynasty with copper printing mould, 33cmX22cm. It is preserved in the National Museum of Chinese History.

3-12 Luo Xuan Bian Gu Jian Pu, 31.5cmX21cm, replica from reproduction of original, block printing without ink, watercolour block printing and block printing in chromatography, dating back to 1626. It is preserved in the National Museum of Chinese History.

3-12

3-13

3-13 *Yi Lin Gai Cuo* , from wood engraved block edition, Qing Dynasty, 27.3cmX16.7cmX1.6cm. The book deals with corrections of errors of medical works and is in two volumes, written by Wang Qingren. It is kept in the National Museum of Chinese History.

3-14

3-14 Seven specimens of watercolour block printing in different sizes: 24cmX16.5cm, 13.8cmX10cm, 14.9cmX10.9cm, 10cmX4.7cm, 16.9cmX9.4cm, 13.7cmX4.7cm and 8.5cmX3.7cm. Watercolour blockprinting began in the Ming Dynasty. The blocks, engraved according to colour, went through chromatography process printing or off print. This block printing reproduces *Shrimps* by the contemporary artist Qi Baishi.

3-15

3-15 *Shrimps* by famous contemporary painter Qi Baishi, printed with watercolour block printing technique.

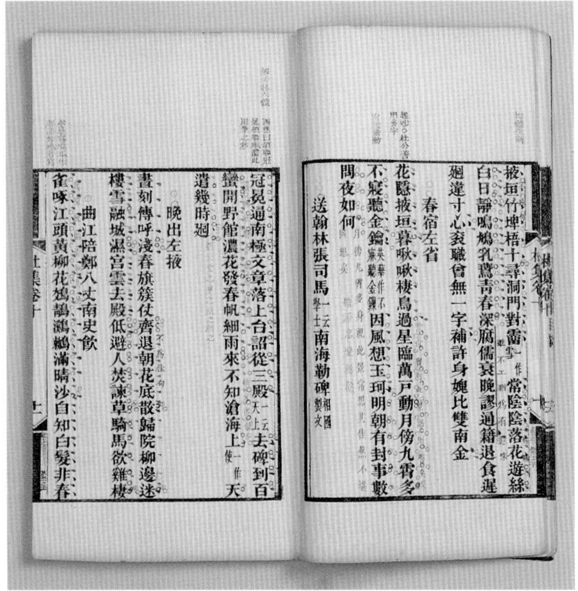
掖垣竹埤梧十尋洞門對霤一作雪常陰陰落花遊絲
白日靜鳴鳩乳鷰青春深腐儒衰晚謬通籍退食遲
廻違寸心袞職曾無一字補許身媿比雙南金

春宿左省

花隱掖垣暮啾啾棲鳥過星臨萬戶動月傍九霄多
不寢聽金鑰英華作不寐聽金鑰因風想玉珂明朝有封事數
問夜如何

送翰林張司馬一云學士南海勒碑相國製文

冠冕通南極文章落上台詔從三殿一云天上去碑到百
蠻開野館濃花發春帆細雨來不知滄海上一作使天
遣幾時廻

晚出左掖

晝刻傳呼淺春旗簇仗齊退朝花底散歸院柳邊迷
樓雪融城濕宮雲去殿低避人焚諫草騎馬欲雞棲

曲江陪鄭八丈南史飲

雀啄江頭黃柳花鵁鶄鸂鶒滿晴沙自知白髮非春

3-16

3-17

3-18*

3-18

3-16 Poems by Du Fu, from the book, titled *Du Gong Bu Ji* published in 1749. It is 28cmX17.5cmX1cm, in six colours and kept in the National Museum of Chinese History. This early edition of poems uses more colours in chromatography process.

3-17 New Year picture announcing good luck for the Spring Festival Qin Dynasty 42cmX54.5cm. Printed in chromatography in Weifang, it is kept in the National Museum of Chinese History.

3-18 Bronze vessel of the Duke of Qin. The 104-word inscription, engraved on pottery mould and cast in bronze, technically paved the way for letter press printing and is kept in the National Museum of Chinese History.

3-19

3-20

3-21

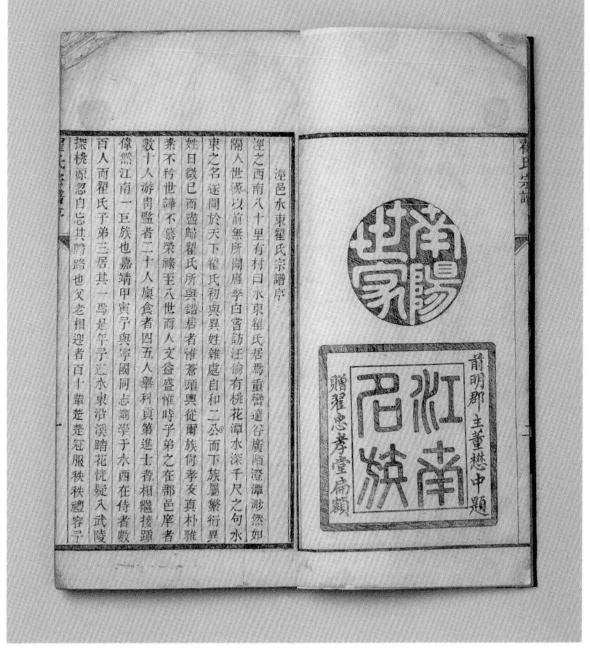

3-22

3-19 Portrait of Bi Sheng, who invented the movable type printing in 1041-1048 in the Song Dynasty.

3-20 Movable type in clay. One iron plate (resin, wax and paper ash) with engraved words is placed above fire, which burns underneath. Another plate is used to press even the engraved words. When cooled the plate can print words. After printing the plate is heated. The two plates are used alternately.

3-21 Movable type specimen by Zhai Jinsheng (Qing Dynasty) who engraved over 100,000 words for printing purpose in five specifications of printed words, based on Bi Sheng method.

3-22 *Zhai Shi Zong Pu* (Pedigree of the Zhai Family), 33.5cmX17.8cm, printed in movable type (clay). It is kept in the National Museum of Chinese History.

3-23*

3-23

3-24

3-25

3-23 *Ji Xiang Bian Zhi Kou He Beng Xu*—Buddhist sutra printed in movable type (wood) in Xi Xia language. A specimen of the earliest letter press extant today, it was excavated in Helan County, Ningxia and kept in the Museum of the Ningxia Hui Autonomous Region.

3-24 Revolving frame on which are placed movable type letters, devised by Wang Zhen. One man read the manuscript while another man revolved frame and took out words for printing from segmented sections. The words were returned to the plate after printing.

3-25 Uygur language letter press in different sizes, Yuan Dynasty. They are 1.4cmX2.6cmX2cm, 1.3cmX0.9cmX2.2cm, 1.4cmX0.5cmX2cm, 1.4cmX0.8cmX2cm and 1.4cmX0.8cmX2cm. Discovered from Mogao Grottoes, Dunhuang, Gansu, it is dated to the Yuan Dynasty (1271-1368).

3-26

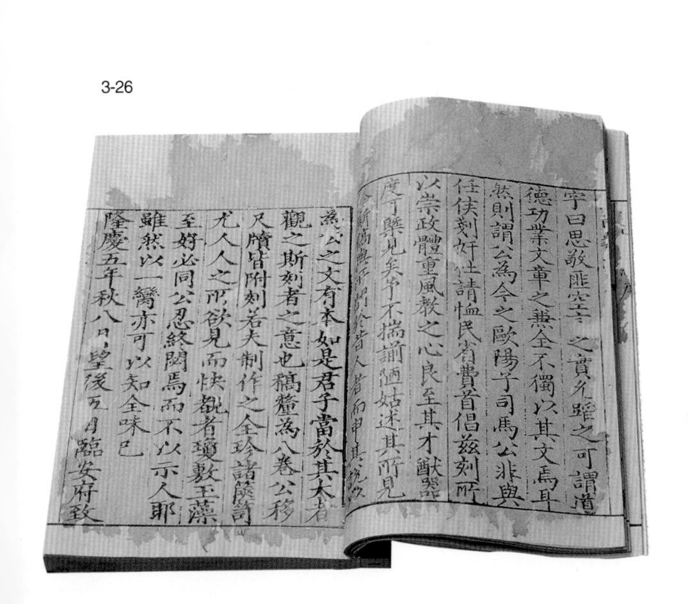

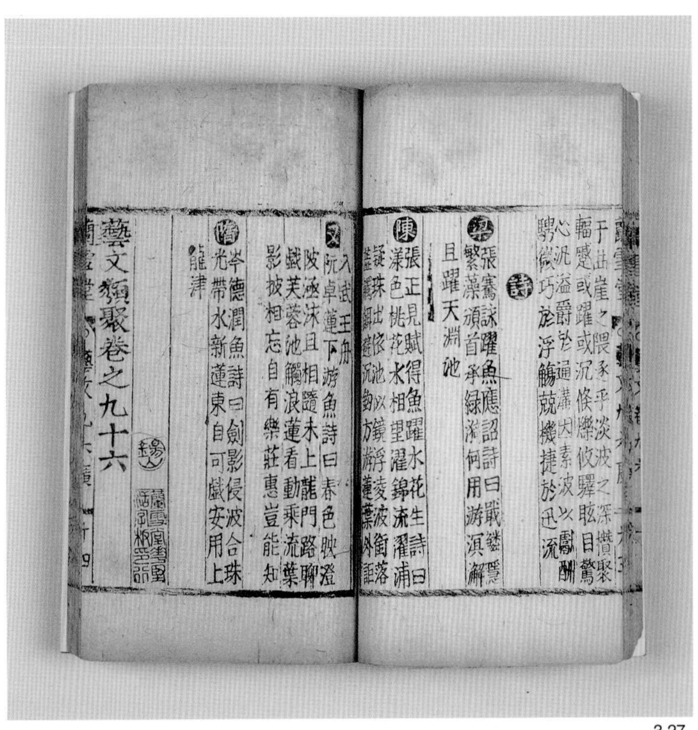

3-27

3-26 Edition of *Qian Nan Lei Bian,* printed in movable type (wood). Measuring 27.2cmX20.3cm, it is kept in the National Museum of Chinese History.

3-27 *Yi Wen Lei Ju,* printed in movable type (copper) made by Mr. Hua of Wuxi. Earliest specimen extant, it measures 22.3cmX14.5cm and is kept in the National Museum of Chinese History.

Chinese Printing Spreads To The World

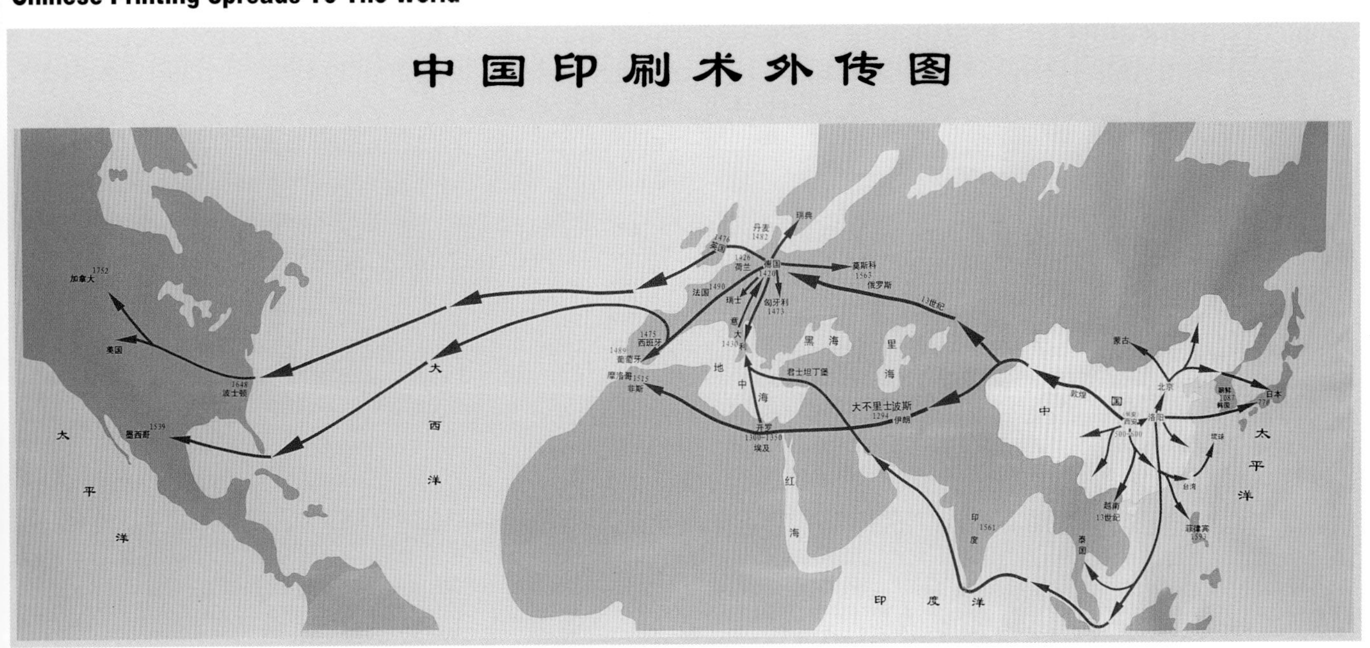

COMPASS

The Chinese are the first to discover the property of the magnet, which points from north to south in direction. Since the south is revered in China, the compass specifically points to the south. It is known as zhi nan zhen (south pointing needle). The course of the development of magnetic tool for pointing direction is divided into two stages.

One. In the Warring States Period (475B.C.-221B.C.) a device known as si nan (south pointer) was made. This is recorded in the book: *Han Fei Zhi: You Du Bian*, which says that si nan was used to "duan zhao xi" —to indicate east, south, west and north. The Chinese already had a clear perception of direction at this time. The Han Dynasty si nan was made of bronze, with an earth plate and magnet. It was circular inside and square on the outer rim. The entire piece was made smooth and glossy as a result of rubbing. Heavenly stems and earthly branches plus four symbolic eight diagrams are arranged alternately. The word zi is due north. Going clockwise the magnetic scoop points to mao(due east), wu (due south) and you (due west).There are 24 locations in the diagram. The magnetic scoop can turn round. When it stands still, the head and tail point from north to south. This compass was used in China for over 1000 years until the Tang Dynasty. Two. In the Northern Song Dynasty (960-1127A.D.) two changes took place: the natural magnet or lodestone was replaced by man-made magnet and the scoop by the magnetic needle. Man-made magnet was produced by electromagnetic induction or friction. In the case of the former an iron substance goes through the process of burning. It is taken out and placed in the position of north and south. Due to induction of earth magnetism, iron becomes magnetic when it is cooled. In the case of the latter rub an iron pin on natural magnet to produce magnetism and the result is man-made magnetism. The invention of man-made magnetism is of crucial importance in the history of the development of magnetism and geomagnetism. In the Northern Song Dynasty water-floating fish magnetic compass was made by using geomagnetism. Other compasses are: fingernail compass, bowl rim compass, suspended method and water-floating compass made by friction, which imparts magnetism. Magnetic needle is placed on fingernail or porcelain bowl rim. The needle easily drops down. It points south and north. This is best for experimental purpose only.

The suspended method uses a silk thread to hang the magnetic needle. A wooden plate with 24 positions is placed below the needle. When the needle stands still, the two ends point south and north. In experimenting needle compass, Shen Kuo discovered that the needle did not point exactly due north or south. There was a slight deviation to the east. He found an angle of declination, forming a wedge between magnetic meridian and real meridian. The wedge moved with the rotation of earth. He took this into consideration in giving accurate direction. His discovery laid a firm foundation for the scientific use of the compass.

The water floating method uses a magnetic needle to go through a wick which is placed in a porcelain bowl. The wick is light and floats above water. The water has a tendency to balance. So long as there is no sudden jerk the needle will point north and south. The method was used in navigation. Zhu Yu of the Northern Song Dynasty in his *Ping Zhou Ke Tan* (Pingzhou Table Talk) says that Chinese merchant ships entering and leaving Guangzhou port were all equipped with compass. Xu Jing in his *Xuan He Feng Shi Gao Li Tu Jing* points out explicitly that water floating compass was used by the ship that took a Chinese envoy to Korea in 1123. Compass was a navigational aid used for all kinds of weather. It opened a new era in navigation history. The development of navigation called for good compass. This in turn prompted the development of the compass. In the Southern Song Dynasty magnetic needle and the device marking location were assembled together to make a complete instrument called pan zhen, jing pan or di luo. The name was changed to luo pan or compass. Pointed by the needle are what is called 24 locations or wedge needle. Actually the locations increased to 48 over the years. Southern Song compasses were of two kinds: water compass and dry compass. A pottery figurine called Zhang the Celestial (unearthed in Zhu Jinan grave, Linchuan, Jiangxi, 1985) holds a dry compass in left hand, using a pointed object to support the needle, as distinct from water floating compass.

The use of magnetic compass in navigation in early days required knowledge of astronomy and geography. Changes in direction of sail by ship and changes in direction of needle on compass were summed up to become the needle route in navigation. With needle route, navigation no longer required astronomical guide. Ocean-going ships only required compass for navigation. Changes in the compass as the needle moved around was called needle locations, which were marked in words. The earliest such record is made by Zhou Daguan of the Yuan Dynasty in *Zhen La Feng Tu Ji* (Records of Cambodia). He sailed to Cambodia in 1295 and kept a record of his compass route. At the start of the book he says: "We set sail to the ocean from Wenzhou port and sailed in the sea outside Fujian. Passing the Qizhouyang, the sea outside Vietnam, we arrived in Zhen Pu (Cambodia) in half a month via Zhancheng with favourable wind." Zheng He, the famous navigator of the Yuan Dynasty, made seven voyages in the ocean. He recorded his navigation route by means of compass needle positions.

The support style compass was successfully made in the Yuan Dynasty in the 14th century. The shape of the compass is like a tortoise (made of wood). The belly contains magnet. The tortoise is placed on a pointed pillar. The centre is the support point. The tortoise turns round. As it comes to a halt, its head and tail point south and north. It is not a modern needle compass. As yet there is no designation of locations on the compass.

The magnetic compass invented by China, spreading to Asia and Europe since 13th century, greatly prompted navigation and promoted economic and cultural exchanges among peoples of the world. All told, the Chinese compass played a big role in the development of economy and culture of the world.

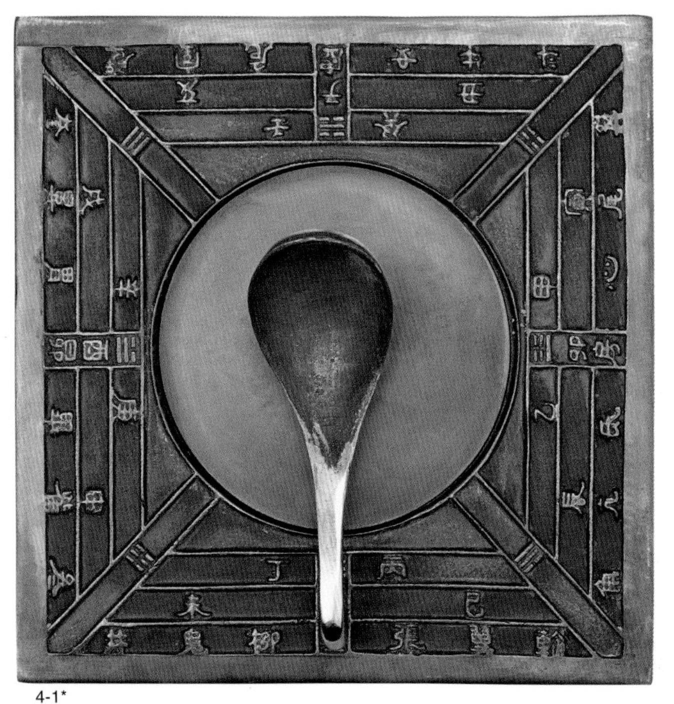
4-1*

4-1

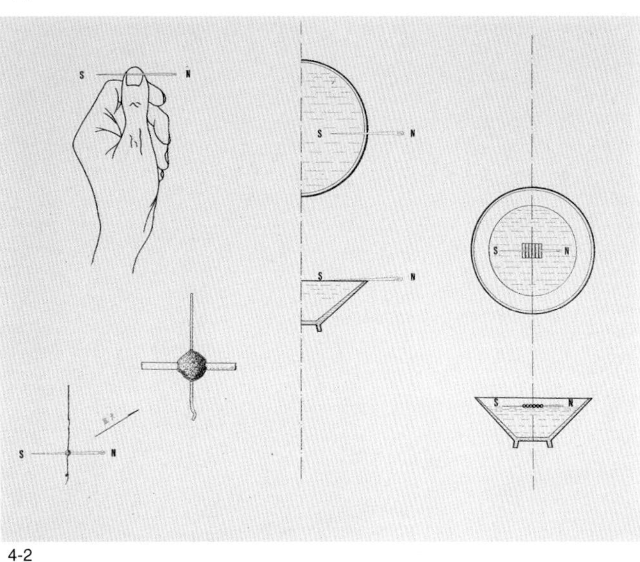
4-2

4-3

4-1 Reproduction of Han Dynasty si nan, 17.8cmX17.4cm. The scoop is 11.5cm long and 4.2cm. in diameter. Made up of a bronze di pan and natural magnetic scoop(head pointing N and tail S), si nan has several tiers on the di pan, depicting 8 heavenly stems, 12 earthly branches, 4 divination symbols and 24 locations. This replica was made by Wang Zhenduo based on a description in *Lun Heng* (Discourses Weighed in the Balance) and on the di pan unearthed from a Han Dynasty grave.

4-2 Diagram of suspended, water floating, fingernail and bowl rim compasses. The two last mentioned compasses have magnetic steel needle on the rim of the bowl or fingernail. The needle uses the touching point as support and swings left and right. When not in motion the two ends of the needle point north and south.

4-3 Model of a Northern Song Dynasty suspended compass, 38cmX21.5cmX21.5cm X21.5cm. A waxed silk thread is placed in the middle of the magnetic needle, suspended on a wooden frame. Underneath is placed a location disc. The two ends of the needle point S and N. The model is made by Wang Zhenduo based on a description in *Meng Xi Bi Tan* (Notes Written in Dreams).

4-4

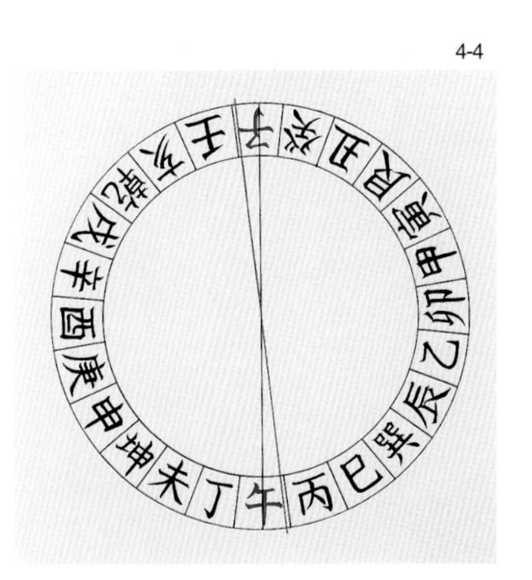

4-4 Slight deviation between terrestial pole and magnetic pole discovered by Shen Kuo, Northern Song Dynasty, is shown in diagram. The deviation angle differs from time to time. The North and South Poles, linked by meridian, are used as the basis for pointing directions. The terrestial pole and magnetic pole do not coincide, however.

4-5 Northern Song water floating compass, 2cm high and 10cm in diameter. The diameter of the bottom is 3.5cm. A segmented wick is appended to the magnetic steel needle in a bowl containing water. The needle and wick float on water, pointing S and N. The compass, with practical value, was used as a navigational guide. The model is made by Wang Zhenduo based on descriptions of the instrument in *Meng Xi Bi Dan* and *Ben Cao Yan Yi.*

4-5

4-6

4-6 Yuan Dynasty bowl compass, 7.5cm high and 17.8cm. in diameter. Unearthed from Ganjingzi, Lushun in 1959, it is kept in the Lushun City Museum, Liaoning Province. The bowl is a white glazed floral design in brown made in Cizhou Kiln, Hebei. The word needle in Chinese is written at the bottom of the bowl. The design shows a floating compass needle. This shows that the bowl was used to contain compass for navigation.

4-6**

4-6*

4-7

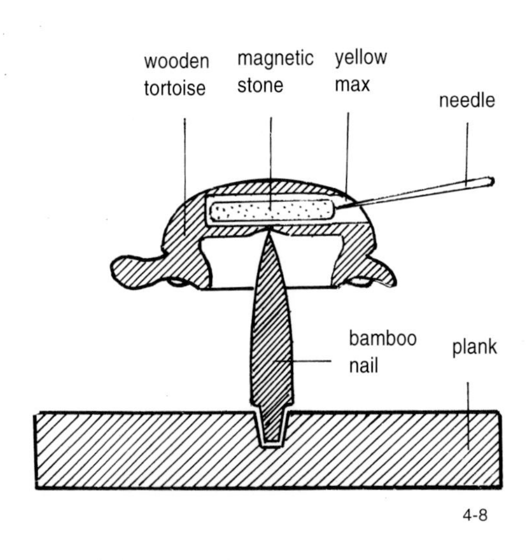

4-8

4-9

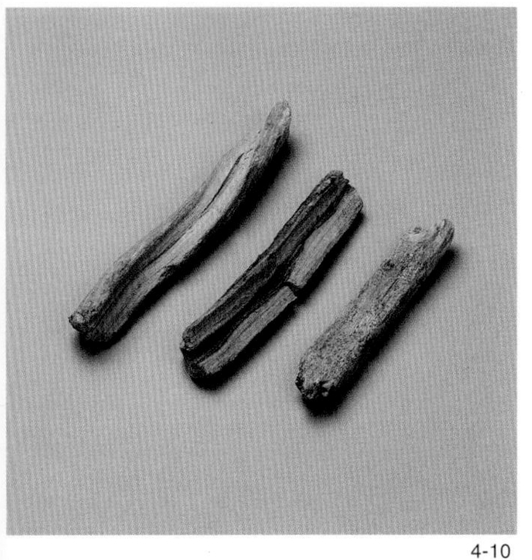

4-10

4-11

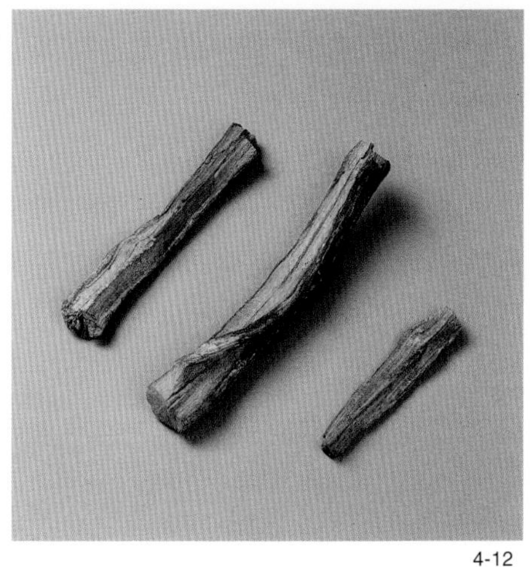

4-12

4-7 Southern Song Dynasty porcelain figurine holding compass in hand, 22.2cm high—an artefact, unearthed from Linchuan grave, Jiangxi and is kept in the Shangdu Jurisdiction Area Cultural Relics Management Bureau. The model of a geomancer, it provides information on the time, shape and work of dry compass.

4-8 Schematic diagram of tortoise compass, Yuan Dynasty. The tortoise compass is a supportive type of magnetic compass, with a length of 15cmX15cm on the edge. The tortoise is 11cm in length and has a height of 13cm. A magnetic stone is stored in the belly. The wooden tortoise is placed on the vertical axle. The head and tail of the tortoise point S and N. The diagram shows the structure of the tortoise.

4-9 Model of sea-going ship, Southern Song Dynasty. There is a keel, with 13 water tight cabins at the bottom, separated by 12 wooden partition walls. Is measures 177cmX60cmX183cm. The original ship was unearthed from Houzhugang, Quanzhou in 1974. It measures 24.2mX9.15m.

4-10 Eaglewood carried by Southern Song Dynasty sea-going ship. It was unearthed from the ship buried in Houzhugang, Quanzhou, now kept in the Quanzhou Museum of History of Overseas Communications.

4-11 Sandalwood unearthed from same ship, kept in the same museum.

4-12 Jiang zhen xiang (acronychia pedunculata)—Chinese spice carried by same ship, Quanzhou, Fujian and preserved in the same museum.

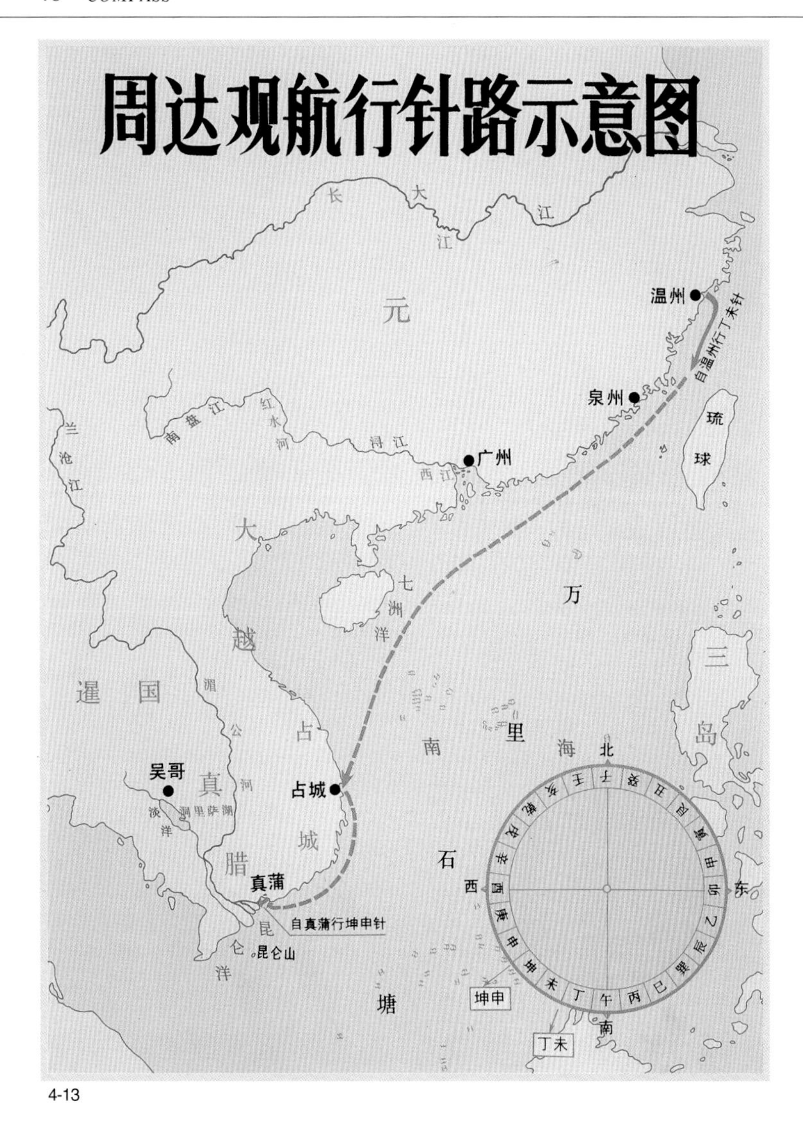

4-13

4-13 Route Indicated on Compass Map by Zhou Daguan

4-14 Model of ship sailed by Zheng He This biggest ship of Zheng He's fleet was 444 ft long, and 180 ft wide. There were 9 mastheads carrying 12 sails. It used astronomical and water floating compass as navigational aid. Zheng He launched seven expeditions 1405-1433. The ship was the biggest sailing boat in the world at the time.

4-15 Navigation route of Zheng He, who sailed from Nanjing to as far as East Africa. The map was used by Zheng He in his last expedition in 1430. The original map was in scroll. The map shows over 500 ports or cities, with landforms of mountains and rivers. The navigation route is indicated by compass with location and voyage provided on the disc.

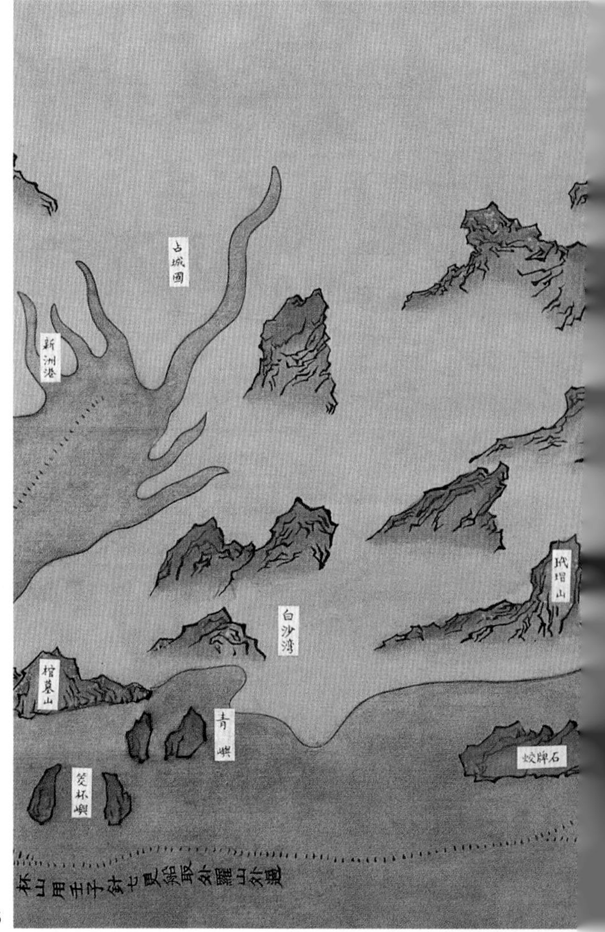

4-15

4-14

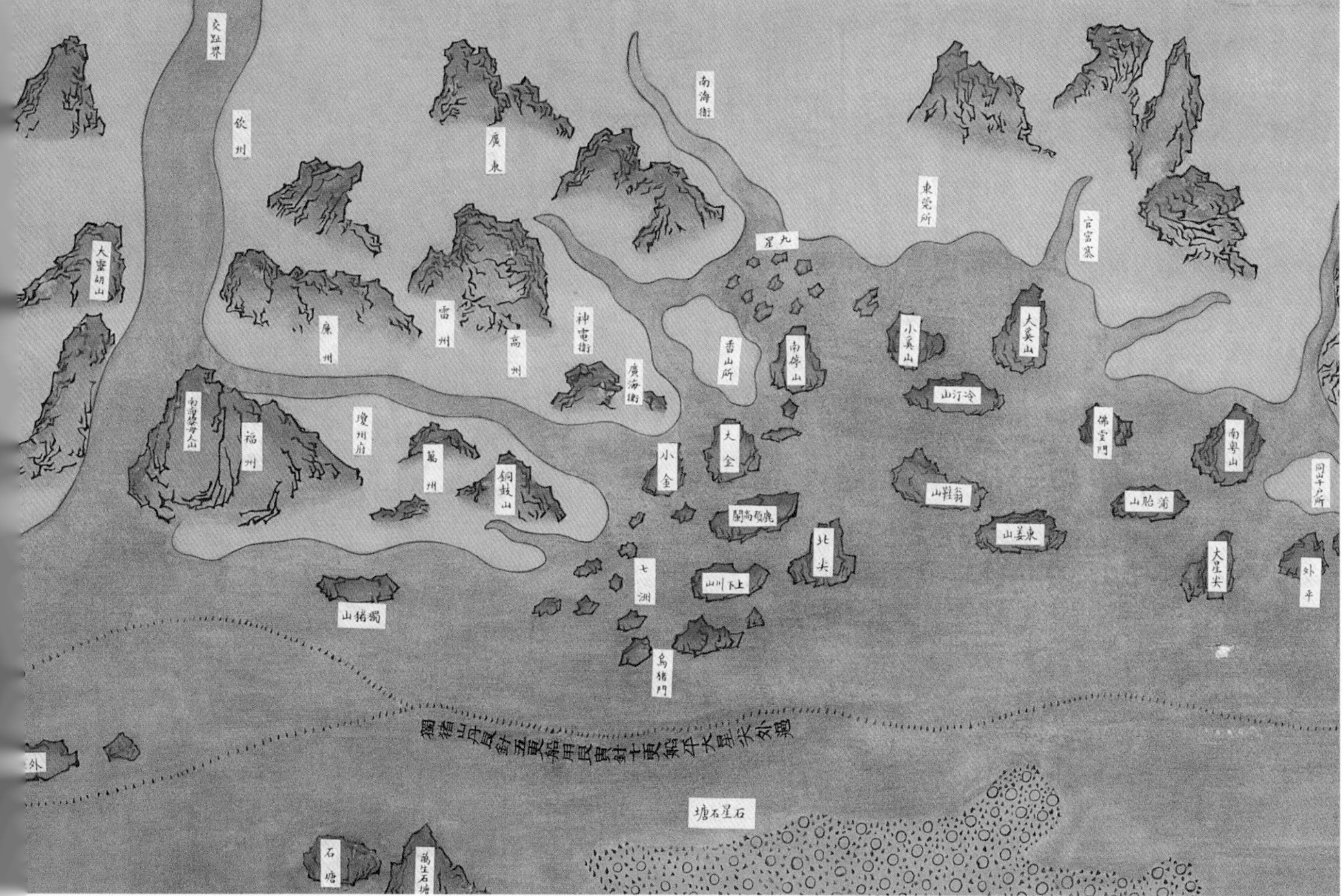

GUNPOWDER

One of four great Chinese inventions, gunpowder has great significance in the history of chemistry in the world. Ancient Chinese gunpowder was made of saltpeter, sulfur and materials containing charcoal in suitable proportions. Saltpeter and sulfur were regarded as medicinal substances in the Han Dynasty. They were used as material in alchemy or pill refining, playing a crucial role in the invention of gunpowder. *Can Tong Qi* (combination of *The Book of Changes, Huang Lao* and *Luo Huo*), a book by Wei Boyang, Eastern Han Dynasty says that when chemicals of like nature are heated they mix well. Otherwise there will be no response. If chemicals of opposite nature are heated, the reaction can be tremendous and abnormal phenomenon will often result. Ge Hong of the Eastern Jin Dynasty makes reference repeatedly to saltpeter, sulfur and realgar in his well-known book: *Bao Po Zi* (The Book of Master Bao Pu). He notes the chemical transformation when saltpeter, resin and carbohydrate are put together and burnt.

Black gunpowder is made of combustion and combustion supporting substances. The former includes sulfur, realgar, resin, charcoal and carbonhydrate. The materials that support combustion are saltpeter and others. In the composition of early gunpowder saltpeter is indispensable. The combustion materials are not fixed. They may be sulfur, realgar or other materials. In the course of development of gunpowder, saltpeter, sulfur and charcoal are regularly used.

Feihuo (literally flying fire, an early term for gunpowder) was formally used as firearm in China no later than 904A.D. The firearms were bombarded at city towers guarding city wall during military operation with the aid of arrows or stone slinging devices. Firepower increased, thanks to the invention of gunpowder, as burning by gunpowder and increased shooting range were combined, inflicting heavy casualties on the enemy. The feihuo became a new Chinese weaponry, as it strengthened the power of destruction of the cold weapon and increased the distance of operation of firearm. It was an outstanding contribution to weapons of war. In the Song Dynasty gunpowder manufacturing technique was greatly bolstered. The book *Wu Jine Zong Yao* (Essentials of the Book of Weaponry), compiled in the 11th century, used the word huoyao (gunpowder or combustive substance) for the first time. It records three ways of preparing huoyao for three different purposes. This testifies to a ripening of the knowhow of firearms. On the basis of this knowhow on the explosion of gunpowder when fired, three kinds of weapons were manufactured: explosive firearms, spraying firearms and tube firearms.

Explosives appeared very early. The same book records thunderbolt explosives (pili huoqiu) and jili huoqiu. Cannons were used in military operations in the Song, Jin and Yuan dynasties. Combustive substances were contained in pottery or raw iron shells. When ignited they blew up, killing the enemy. Zhen Tian Lei appeared. The noise caused by Zhen Tian Lei in explosion was like a thunder. The fire penetrated into armour.

In the attack on Jingjiang city at the end of the Southern Song Dynasty, defenders fired a cannon, causing a heavy explosion like thunder, destroying the city wall, and killing both the Song defending army and Yuan attackers. In the Ming Dynasty(1368-1644) land mines using gearwheel control device and time torpedoes with joss sticks were used.

The earliest tube firing gun began with the Southern Song Dynasty. The first ever gun with barrel was made in 1259, known as tu huo qiang. A bamboo tube contained the combustive material or bullet. Historians of the world pay great attention to this gun, to which acknowledgement is made as the first gun based on the principle of projection. It is the origin of firearms of the world. Towards the end of the 13th century metal was used for shell casing instead of bamboo. This increased firepower still more. It is the first time that metal casing was used. As a result of this the term huo tong came to be used. In the Yuan Dynasty a ratio of 60%, 20% and 20% was used for saltpeter, sulfur and charcoal. Compared with Song Dynasty weapons the content of saltpeter had obviously increased. The gun consisted of three parts, bore or interior of gun barrel, the gunpowder chamber and tailstock. The ratio between the three parts reflect a quantitative relationship to suit the need of projection. Huo tong were of two kinds: small arms used by soldiers with wooden handle. This developed into modern small arms. The cannon was placed on wooden racks. It had a big diameter. A copper cannon was made in China in 1332. This earliest artefact is now kept in the National Museum of Chinese History, Beijing. The huo tong developed in greater variety and bigger size as excellent weaponry in combats at sea and on land in the Ming and Qing dynasties.

The spray gun used the powder barrel that sprayed backward in launching. When ignited the gas escaped through a rear vent and drove the container forward by the principle of reaction. It did not use arrows. Instead a small rocket was placed in front of the arrow. Ming Dynasty rockets improved. As many as 10 rockets were fired, called Yiwofeng (resembling a bee hive). It was placed on a single-wheel vehicle, called firing chariot. The firing chariot and soldiers equipped with shields were good for both offence and defence. They were called tiger heads. Two-stage rockets, invented in the Ming Dynasty, were the earliest multiple-stage rockets in the world. But the combustive substances were not used for fuelling or explosive. Instead they were used as driving force. Another spray weapon used the same combustive substance as driving force. The combustive substance fired became the bullet. The most representative of this weapon was shen huo fei ya (literally celestial fire flying crow).

Apart from firearms, gunpowder can be used as fireworks display and firecrackers for celebration purposes or for hunting. Gunpowder is used as medicine. Ancient Chinese medical books record the use of gunpowder to cure scab, as insecticide or to end plague. Firecrackers began with the Northern Song Dynasty. Fireworks display registered a peak period in the Ming and Qing dynasties, when they were widely used by the people. There were a wide range of varieties--over several dozens. The more famous fireworks are called jiazideng, hezideng, didijin, qianzhangju and dilaoshu. Saltpeter as medicine spread from China to Arabian countries in the 8th and 9th centuries. The Arabs called them Chinese snow. After the 13th century gunpowder, fireworks and gunpowder weaponry spread to the region. The Arabs called them China flowers or China fire guns. In the 14th century gunpowder spread to Europe from China. This caused an important change in weaponry manufacture, tactics and strategy. Great indeed was the impact on the economy and society of Europe, as a result of the spread of Chinese gunpowder.

5-1

5-2

5-3

5-1 Traditional Chinese painting on pill refining that played a crucial role in gunpowder invention. This is a Chinese traditional painting which illustrates an explosion during pill refining. Many chemicals were used, including sulfur, realgar and sulfate and saltpeter. There was danger of explosion, to which caution was exercised, according to *Can tong Qi.*

5-2 Specimens of saltpeter, sulfur and charcoal used in gunpowder manufacture. Gunpowder used the three elements, whose chemical formula is: 4KNO3+S2+6C=2k2s+2N2+CO2. After heating gas is produced, boosting the volume, which leads to explosion. By the Song Dynasty the Chinese chemists had mastered the technique of making gunpowder.

5-3 Three prescriptions for making gunpowder used in the Northern Song Dynasty. The book *Wu Jing Zong Yao* gives the prescriptions for toxic and jili huoqiu and cannon.

type	gunpowder ingredients			
	saltpeter	sulfur	charcoal powder	other combustible
duyao yanqiu	30%	15%	5%	50%
jili huoqiu	40%	20%	5%	35%
cannon	40%	14%	14%	32%

5-4

5-4 Tube firingarm—tu huo qiang, range 230 meters, described as the first rifle in the world. It was invented by the Song government army. Bamboo tube was used to contain gunpowder and bullets. When the fuse is ignited, the gunpowder explodes and sprays.

5-5

5-5 Yuan Dynasty copper cannon, 35.3cm long, diameter 10.5cm, weighing 6.94 kg. the diameter of the tail end is 7.7cm. Discovered in Yun Ju Si, Fangshan, Beijing, it is kept in the National Museum of Chinese History. It was made in 1332, the earliest huo tong extant.

5-6 Model of Yuan Dynasty rocket. Measuring 108cm, it has a powder tube at the front. When ignited hot air is produced and pushes out of the back of the rocket. This is the earliest rocket in the world.

5-6

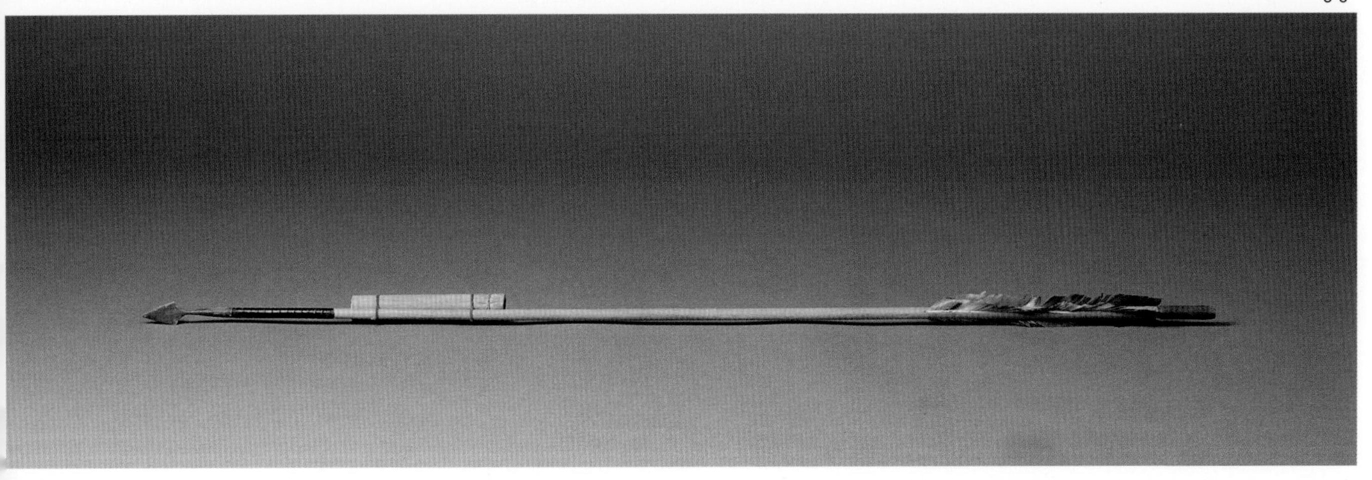

5-7 Ming Dynasty Yiwofeng spray firearm—model. It measures 171cm long, with dozens of rockets connected by thread. The largest diameter is 48.5cm. When ignited, dozens of rockets are fired on the enemy. They look like a bee hive, hence the Chinese name Yiwofeng.

5-8 Ming Dynasty fire chariot, 120cmX52cmX87cm, manned by two persons. It has 160 rockets in 6 tubes, two short guns and two long guns. The vehicle has only one wheel.

5-7

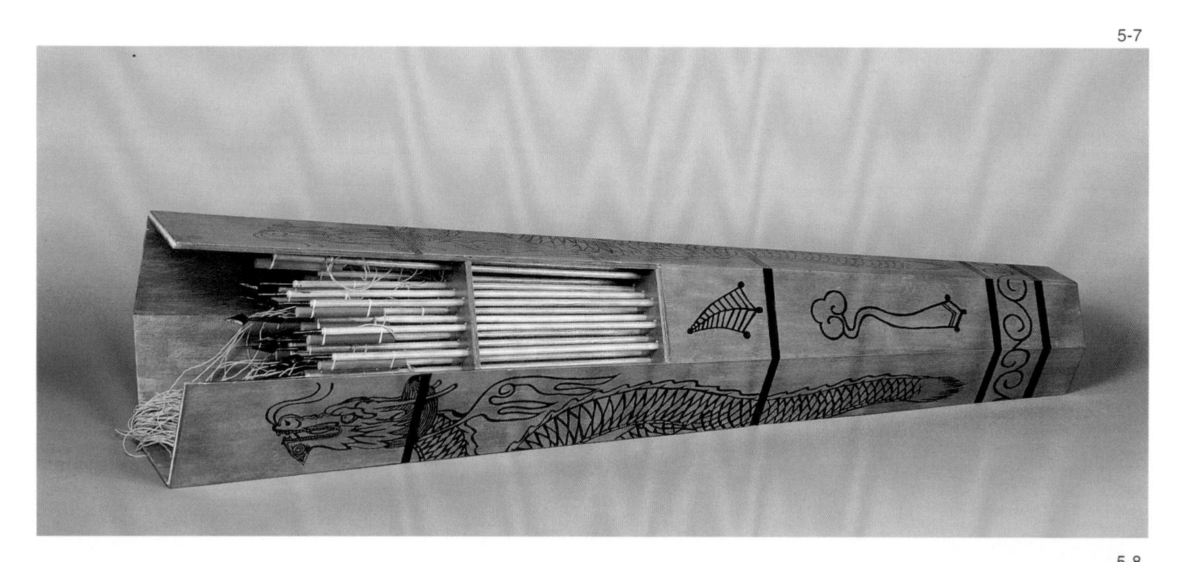

5-8

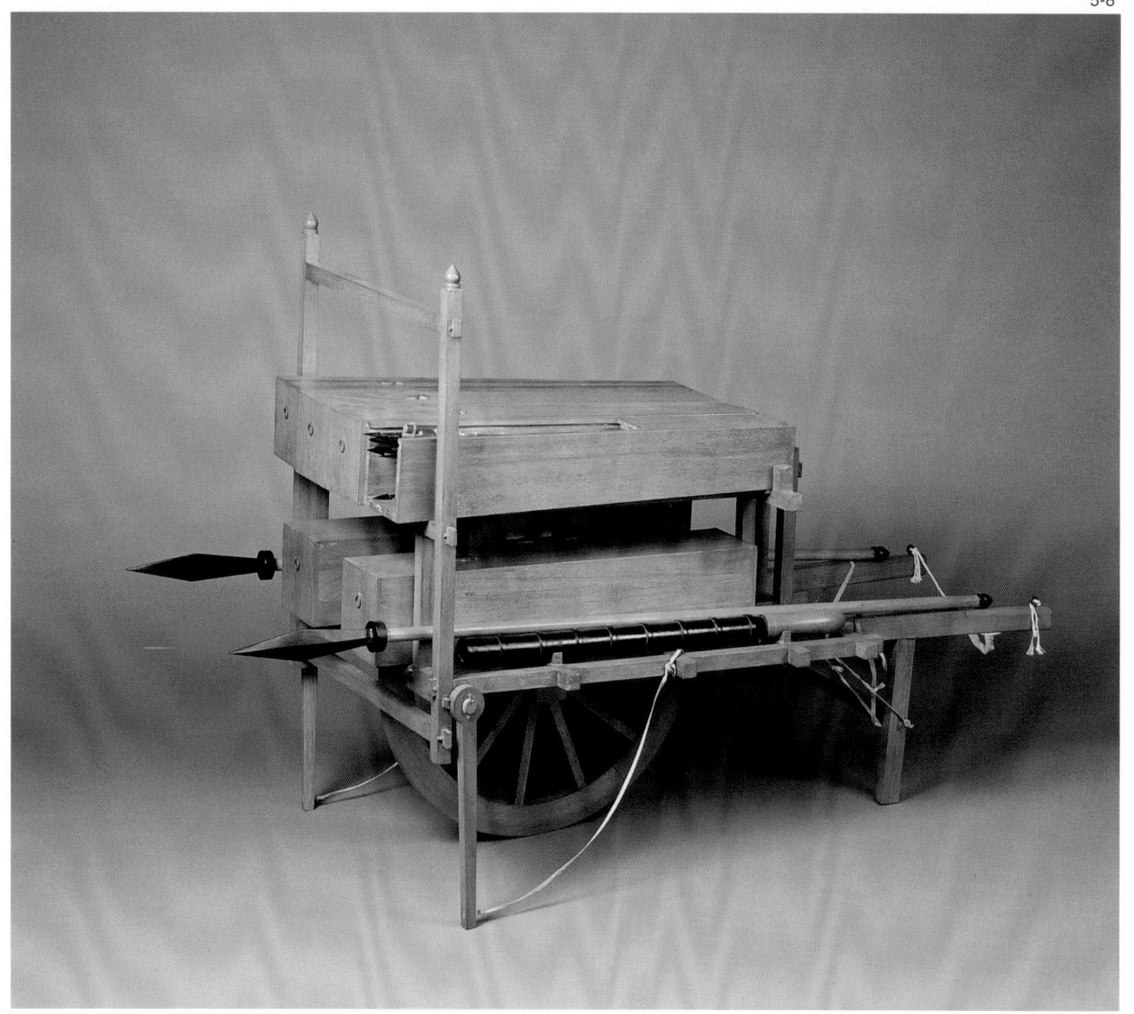

5-9 Model of rocket with combustive substance, earliest 2-stage rocket in the world. It measures 153cm long. The diameter of the head is 20cm. The tail is 32.5cm wide. The body is five feet. Gunpowder tubes are attached to the head and tail, pushing the body forward. The belly is installed with rocket. When the dragon body flies over a distance, the rockets in the belly are ignited and hit the enemy position. The Chinese name for the rocket is fire dragon coming out of the water, as it is fired from the ship.

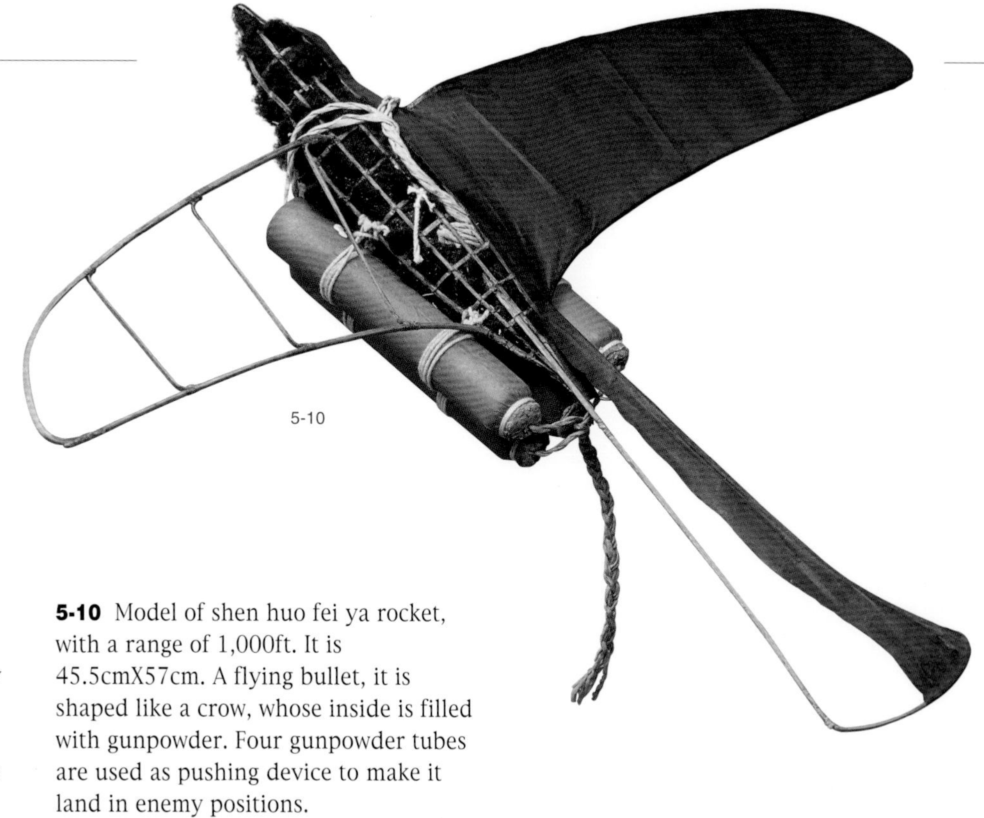

5-10

5-10 Model of shen huo fei ya rocket, with a range of 1,000ft. It is 45.5cmX57cm. A flying bullet, it is shaped like a crow, whose inside is filled with gunpowder. Four gunpowder tubes are used as pushing device to make it land in enemy positions.

5-9

5-11

5-11 Picture showing fireworks display in Ming palace in 1475. Its vertical width is 37cm while its horizontal length is 623.5cm. It is preserved in the National Museum of Chinese History. It depicts a scene in which people in the palace of the Forbidden City enjoy a festival with fireworks display. The year coincided with the 11th reigning year of Emperor Xianzong.

5-12

5-12 Fireworks mounted on framework, watched by crowds in the Ming Dynasty. This is taken from the novel *Jing Ping Mei,* which depicts the life in Chinese metropolis during the Ming dynasty. When ignited the whole fireworks go off, giving a spectacular view.

Ancient Chinese Science and Technology

AGRICULTURE

One of four great nations of the world with ancient agriculture, China established herself as an agricultural nation with a long farming tradition since very early times. Her achievements in ancient agricultural science and technology are magnificent.

Chinese forebears produced many varieties of cereals. Recent archaeological findings in Hemudu, Zhejiang and Li County, Hunan Province reveal that rice seedlings unearthed in graves in the two places were not of the wild variety but strains cultivated over 8,000 years ago. This testifies to the fact that China is one of the sources of rice cultivation in the world. In the Yellow River Valley corn was cultivated more than 7,000 years ago. The world credits China with the earliest cultivation of corn, from which she developed herself as the earliest agricultural state of the world. The successful cultivation of this cereal spread in early times from China to Russia, Austria and the rest of Europe. Corn cultivation from China spread elsewhere via Korea and Japan. In addition to rice and corn, millet and meizi (a variety of millet) were known to have been cultivated in China as is shown by excavations from 7,000-year old graves along the Yellow River Valley. We see that the role China played in the cultivation of cereals had far reaching significance on the development of world agriculture.

Many primitive farm tools going back 7,000-8,000 years ago have been excavated in graves across the vast expanse of China. They include stone axe, stone shovel, wooden and bone shovels with handles for cultivation in virgin land. There are stone knife, pottery knife, stone sickle and sickle made from clam shells, used for harvesting crops. Deserving particular attention is weeder for intertillage or field management. This signifies the fact that field management was introduced then--great progress by Chinese farming in primitive time. The tools are found in great quantity in graves in the middle and lower reaches of the Yellow and Yangtze rivers. The distribution of primitive farming area was very wide. The tools, as we have said, were made of different materials: stone, wood, pottery, bone and shells. In the ruins of human habitation 7,000-8,000 years ago we find buhr or hirst, millstone, mortar, club, pestle—all made of stone. The processing of cereals into fine foodstuff (including grinding and unhusking) was made possible by these tools, which transformed farming into a prosperous industry with substantial increases in output.

The emergence of ox as draft animal to till land and farming tools made of iron spelt a great leap forward in agriculture in China. Inscriptions on tortoise backs and oracle bones of over 3,000 years ago refer to li niu or ploughing ox. This shows the long history of ox in China. The Spring and Autumn provided a more explicit account of this. The book: *Guo Yu: Jin Yu* (Records of the States: Records of Jin) says that the ox, driven by the farmer, was used to till land owned by ancestral temples during this time. In the Warring States Period employing ox to plough land became a popular pursuit, marking a step forward in agriculture and relieving man of the heavy burden of using hands to drive the plough in tilling land.

Iron was referred to as wujin in Chinese history books. The making of iron tools began in the Spring and Autumn Period. By the Warring States Period iron tools were used extensively in China. Excavated in graves in Xinglong, Hebei Province are moulds, in which iron was melted and cast into tools of various shapes: iron hoe, iron axe, and earth digging tool made of the same metal. Sites of iron smelting operation and relics of iron farming tools have recently been unearthed in a great many places: Jilin in the north, Guangdong and Guangxi in the south, Shandong in the east and Sichuan and Gansu in the west. The tools include spade, cha (earth digging tool), hua (plough), grass weeding implement and iron sickle. The manufacture and extensive use of iron tools prompted the development of agriculture in China, for deep ploughing and intensive cultivation were guaranteed by precision drilling implements. This enhanced the speed of harvesting and quality of cereals produced for consumption. In a word, production efficiency was raised to a new high.

The development of agriculture in the Han Dynasty was further enhanced with the introduction of new cultivation technique and better tools. In the reign of Emperor Wudi, Gu introduced the dai tian fa of cultivation. Grooves and ridges were dug. Seeds were sown. When leaves grew on young plants, the farmers weeded the grass. The seedlings were covered with earth. The grooves were piled up with earth. Planting was carried out in new grooves the following year. This method proved good for rationall utilizing the fertility of the soil and effectively protecting the seedlings against drought.

The shape of the plough of the Han Dynasty facilitated farming. A v-shaped iron hood was attached, which could turn round in any direction, maintaining a sharp edge during the operation. Ploughing can be seen in wall pictures and stone engravings of this period. The seed plough (animal-drawn or operated by man) could plough three rows at a time, faster than by humans. The dui and grinding stone became common tools of the people. The dui was operated by man who either pedalled the framework with feet or drove it with hydraulic power. The shape of the implement can be seen in relics of pottery grinding stone or millstone unearthed in Han Dynasty graves. By the Wei and Jin dynasties grinding stone or millstone increased in number—operated by ox or water power. Water -generated paddy pounder was a new addition to existing tools, widely used by people in areas rich in water resources. Large mills were used in the Tang Dynasty. The nona rice mill driven by water power appeared in the Song Dynasty. Similar devices were also used in the Jin and Yuan dynasties. The famous Dunxi Mill has been preserved from the Ming Dynasty to this day. It combines grinding, unhusking and pounding in an integrated manner.

Long years of farming practice had helped the Chinese to know the importance of water to agriculture. The development of agriculture gained momentum in the Warring States Period, when water conservancy and irrigation were launched. The three water conservancy projects were the Ximen Bao Canal, which used the waters of the Zhangshui for irrigation in Cixian, Hebei Province, Zhengguo Canal in Jingyang, Shaanxi Province which linked up the Jingshui with the Luoshui for irrigation, and the famous Dujiangyan in Guanxian, Sichuan Province. The last was built by Li Bing, magistrate of Qinshu commandery, irrigating as many as one million mu of cropland on the West Sichuan Plain. The project, repaired by succeeding dynasties over 2,000 years, is so successful that it benefits farmers to this day.

Adept in utilising surface water, the Chinese are good in developing underground water as well. Way back in the Xia Dynasty the people of this great nation dug wells to get drinking water so that they could live in areas that did not lie immediately alongside rivers and waterways. The digging of wells enabled the Chinese to expand their habitat to a much wider area than usual. This was quite an innovation made by the Chinese. Deep wells dug in allocated areas in the Han Dynasty and connected by underground channels, accumulated water. The people got drinking water as well as water to irrigate cropland. Well digging became popular in Northwest China, inhabited by Chinese minority people. Called karez, the underground channels effectively prevented water from evaporation or leakage. Such wells are still an important means of water conservancy in Northwest China to this day.

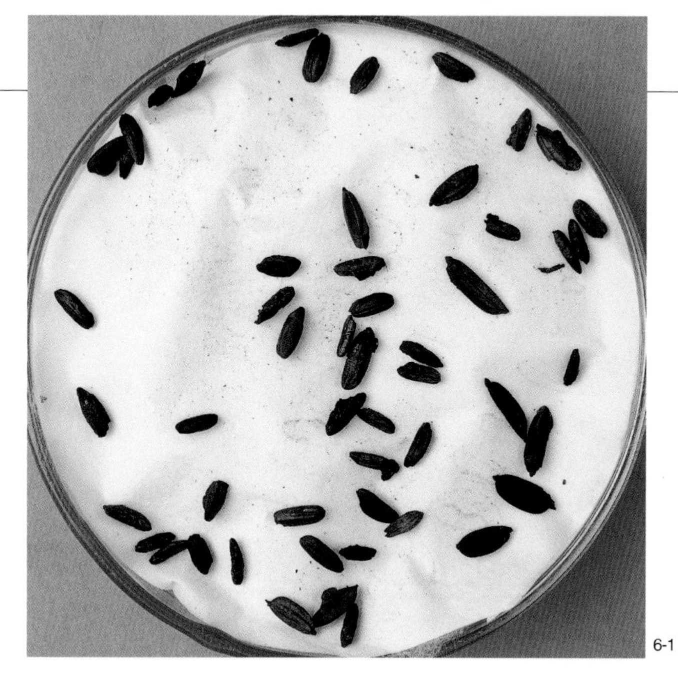

6-1

6-2

6-3

6-1 Rice grains of the New Stone Age unearthed in Bashidang, Lixian, Henan Province. They are kept in the Hunan Provincial Cultural Relics and Archaeology Institute. As the source of origin, China has a history of rice cultivation extending over 8,000 years.

6-2 Corn ashes of the New Stone Age unearthed in Cishan, Hebei Province. They are kept in the Hebei Provincial Cultural Relics and Archaeology Institute.

6-4

6-3 Sorghum grains of over 7,000 years ago, unearthed in Dadiwan, Gansu Province. Sorghum was originally a wild variety. It was cultivated as a cereal by Chinese forebears over 7,000 years ago.

6-4 Stone spade, 30.3cmX10cm, unearthed in Peiligang, Xinzheng, Henan Province. This primitive tool was used to plough wasteland and turn up soil. It is kept in the National Museum of Chinese History.

6-5

6-6

6-5 Shi, a farming tool, unearthed in Hemudu, Yuyao, Zhejiang Province in 1970. Measuring 16cmX10cm, this primitive tool is made of bone, which testifies to the wide range of raw materials used in making farm tools. It is preserved in the National Museum of Chinese History.

6-6 Drawing showing a man digging soil with shi.

6-7 Stone weeding implement of the New Stone Age, unearthed in Qiansanyang, Wuxing, Zhejiang. Measuring 18cmX6.5cm, it is preserved in the National Museum of Chinese History. This tool later evolved into the plough. The artefact was a ritual implement modelled on the practical tool. It shows that intertillage was carried out at this early stage.

6-8 Diagram showing the use of the stone weeder.

6-7

6-8

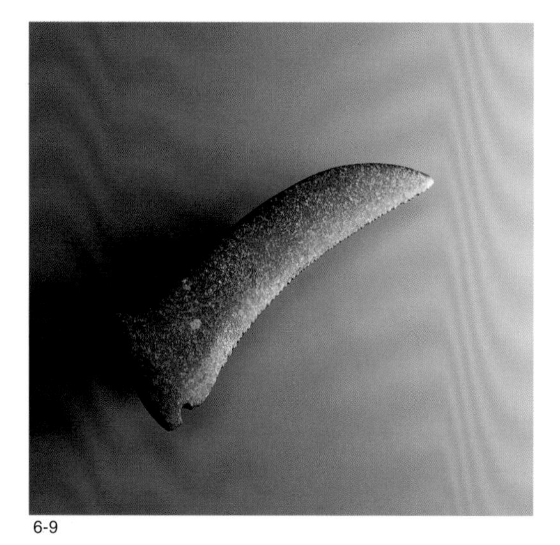

6-9

6-10

6-11

6-11 Warring States Period iron plough, 10cmX10.5cm unearthed in Huixian, Henan in 1950. It is kept in the National Museum of Chinese History. It has a handle and is easy to use. It weeds grass and preserves moisture in the soil.

6-12 Improved iron plough, 23.6cm wide, the blade is 17.5-18cm long and the mid blade is 3.9-4.5cm. The Warring States Period tool was unearthed in Huixian, Henan in 1950. The use of the iron plough created the condition for intensive farming. An embodiment of progress, it raised productive efficiency and was favorable to reclamation. It is kept in the National Museum of Chinese History.

6-9 Seesaw stone sickle of the New Stone Age, 20.6cmX6cm, unearthed in Jiaxian, Henan. This harvesting tool shows progress was being made in farming. The use of the tool boosted harvesting speed. It enabled stalks to be gathered and used. It is preserved in the National Museum of Chinese History.

6-10 Neolithic stone grinding tool for unhusking cereals. The mortar measures 7.3cmX63.5cmX28cm. The stick is 47cm long, 4.8cm in diameter, unearthed in Peiligang, Xinzheng, Henan in 1978.

6-12

6-13

6-13 Han Dynasty tool with seed box and three planting devices, drawn by man or animal. This seed box is a model based on the descriptions provided by the *Nong Shu* (Book on Agriculture) by Wang Zhen, a wall painting on the device in a Han Dynasty tomb unearthed in Pinglu, Shanxi, and references on early planting device in the Nanyang region.

6-14 Water-powered pounding tool (dui) to unhusk rice. This model is made, on the basis of information from two books: *Huan Tan Xin Lun* and *Nong Shu.* The dui first appeared in the Western Han Dynasty (206BC-8AD).

6-14

6-15

6-15 Pottery grinding tool, a burial object unearthed from a Western Han grave, Luoyang, Henan in 1953. It is kept in the National Museum of Chinese History.

6-16 Water-powered paddy pounder (model), made by Cui Liang, Northern Wei Dynasty (386-534). It was mainly used to unhusk cereals. The model is made on the basis of information from the "Biography of Cui Liang" in *Wei Shu* or *History of the Northern Wei Dynasty,* and *Nong Shu* by Wang Zhen. The pounder used water power to drive the horizontal wheel which pushed the stone grinder to crush cereals into powder.

6-17 Model of water-powered grinder made by Du Yu, Western Jin Dynasty. It used water power to drive the horizontal wheel and pushed the stone grinder to crush cereals into powder. The model is made on the basis of information from the "Biography of Cui Liang" in *Wei Shu* or *History of the Northern Wei Dynasty,* and *Nong Shu* by Wang Zhen.

6-16

6-17

6-18 Dujiangyan Dam (sand table). The dam, built by Li Bing in 256B.C. still irrigates cropland to this day. It has a fish mouth to divide waters and regulate water flow, a flying sand dam and treasury bottle mouth, which is the head of the dam. It guides waters from the Minjiang River to irrigate the Chengdu Plain.

6-19 Tashanyan Dam near Ningbo, Zhejiang was built in the Tang Dynasty (618-907). As a dam, it prevented tidal water from flooding rice fields. The water in the reservoir was used to irrigate land. It played an important role in the development of East Zhejiang agriculture and was a famous water conservancy project of ancient China.

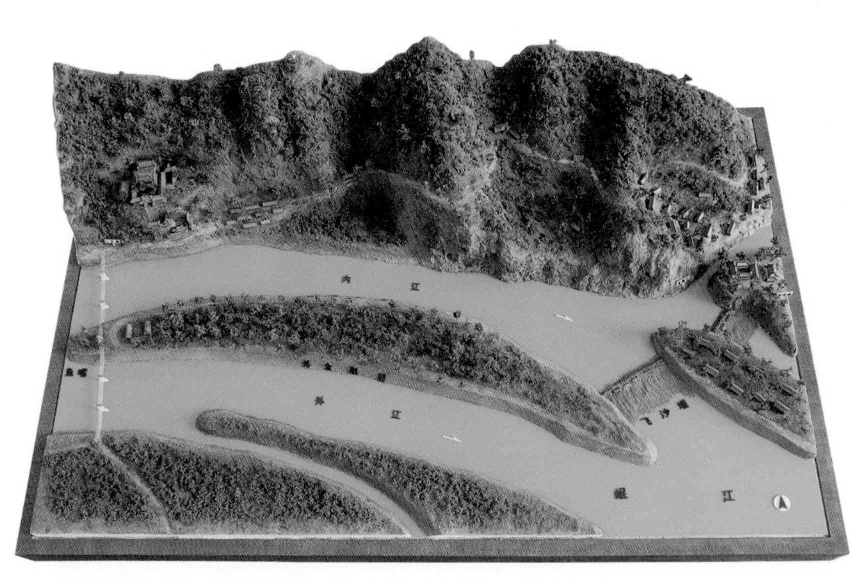

6-18

6-19

CERAMICS

Ceramics is an activity pursued by man to make a natural substance (clay) into a new substance (pottery) by means of chemical process (baking). As early as 10,000 years ago, Chinese forebears began producing ceramics. Ceramic artefacts on the era have been unearthed in Wannian Xianrendong, Jiangxi Province and Xushui, Hebei Province.

In early and middle New Stone Age the hand was generally used to shape clay into desired objects by kneading or pasting design on the clay, coil method and mould. Red pottery and painted pottery became prevalent in the Yangshao Culture. Thin-necked pots with fish or bird design and painted pottery with deer design are representative works of the middle phase of the New Stone Age. In late New Stone Age the wheel method of making pottery was invented to make pottery objects. The potter's wheel is a disc workstation, standing upright and revolving on an axle.

The biscuits or paste are put in the central part of the wheel. As the wheel revolves the potter kneads with hand or tool to make the clay into desired shape with a smooth surface. The ceramic has even sidewall in paste and is beautiful in appearance. The efficiency of pottery industry is raised with the help of the wheel. The wheel-made pottery retains parallel tread on sidewall in concentrated form. The bottom shows eccentric tread cut with thread. Relics of this kind of pottery unearthed in the Longshan Culture sites are monochromatic, without paintings of any kind. They are mostly black. The paste is as thin as an eggshell, hence the name: eggshell pottery. They represent a maximum level of pottery manufacture of this period. Relics unearthed in Miaodigou, Shaanxian, Henan Province of the early Longshan Culture reveal pottery with a contracted upper part, which makes it easy to seal. There are 25 orifices at the bottom. Judging from the composition of the pottery it is clear that the potter must have been successful in bringing temperature under control during the firing or baking process.

In the Shang Dynasty a breakthrough was made in ceramic manufacture with the emergence of stamped hard ware and proto -porcelain. The former required higher temperature in firing than ordinary porcelain. Its paste usually was grey in colour. The latter was made of kaolin. Firing called for a temperature of 1,200 degrees or above. The quality of porcelain was very tough, absorbing little water. Grey and blue glazes were applied on the proto-porcelain, which possessed the basic features of porcelain. The paste and glaze were inferior and temperature not as high as products of later age. It was transitory and primitive in nature.

In the wake of over 1,000 years of development from the Shang to the Western Han Dynasty this ancicnt form of green glaze greatly improved in quality and quantity. It was characterised by hard paste, thick layer of glaze, deeper colours—usually blue-green or brownish yellow (tawny). By the Eastern Han Dynasty celadon consummated in shape and firing skill. The most famous celadon was produced in the eastern parts of Zhejiang Province. The Long Kiln made pale green pottery glaze (finer paste and more glossy). Both the paste and glaze were better integrated. The quality rose to a new high.

In over 300 years from the Three Kingdoms to the Southern -Northern Dynasties celadon entered a crucial stage of development in China. The ceramic industry extended from the South to the North and the Southern and Northern Ceramics Schools were formed. Output and variety rapidly increased. Decorative designs became numerous. Apart from green there were the following glazes: yellow, brown (soy sauce colour), black and tawny. In early 6th century a white porcelain was successfully made in North China, marking a strikingly important advance in porcelain technology.

Blanc de china , as it is called in French (meaning the white of China) contains less than 1 per cent iron in paste and glaze. Oxidization has to be handled well in firing or baking. The blanc de china paved the way for the blue and white porcelain, red within the glaze (or underglaze red), wu cai (five coloured) dou cai (contending colours or strongly contrasting colours) and soft colours or famille verte (Chinese porcelain with designs in which respectively pink or green was prominent). In kilns of this period saggar (clay box was used to pack pottery for baking. The paste was kept away from open fire during firing. The saggar was used by potters of later generations—a legacy from the Shang Dynasty.

The Sui, Tang and Five Dynasties marked an important turning point in Chinese porcelain industry, as production expanded in scale and new technology and style emerged. The Yue Kiln celadon and the white porcelain of the Xing Kiln represented the highest baking technique. The celadon was marked by fine quality of paste and glossiness in colour, described as lei yue, lei bing or qian feng cui se (as beautiful as jade, ice or green peak). Of these the olive green Yue wares were articles of tiptop quality. The olive green porcelains with silver lid or gold-silver ornaments have been unearthed in the Fa Men Si Temple. They are truly gorgeous. The Xing Kiln vessels were white as silver, pure and highly refined. The three-coloured glazed porcelain showed the style of the peak prosperity era of the Tang Dynasty culture —magnificent and in fine shape. There were yellow green, black green and floral glaze made by a process of whipping paste, something quite new in technology. In late Tang Dynasty the Changsha Kiln had a new baking method and produced coloured underglaze, which was unique and laid the foundation for famous Song Dynasty porcelain.

The Song and Yuan dynasties were most remarkable for their achievements in the manufacture of ceramics as visual art—shaping clay into objects of different shapes for aesthetic purpose combined with utility or practical use. Ceramic objects of great beauty were produced. Apart from government-operated kilns many private or civilian-operated kilns came into existence. Song craftsmen, having mastered the technique of glazing and controlling temperature, put out porcelains in various colours. The new wares were Yaobian or kiln transmutation and Kaipian with colors like rose and agate. Song Dynasty kilns had their respective qualities. Jun Kiln wares contained rouge red and turquoise colour. The Ri Kiln wares chiefly consisted of green glaze, crystal like and glossy. The Ge Kiln wares had water finished colours, crystal like and sleek, supplemented with Kaipian. Government kilns produced translucent objects. The best ones were light greenish blue, moon white (clair-de-lune) and blue glaze. The Ding Kiln was famous for its white porcelains, supplemented with engraved design, underglaze relief and stamped decoration. They formed a unique porcelain. In addition to the five government kilns there were the Longquan, Jingdezhen, Yaozhou, Jianyang, Dehua, Cizhou and Chaozhou kilns, each famous for white, blue or green, glaze, design or distinctive type.

The Song Dynasty made much improvement in composition or structure of the kilns and in tools. The Jun, Yaozhou and Longquan Kilns were most representative of the kiln design. All the kilns had saggar or clay box, in which pottery was packed for baking. The paste got heated in an even manner, avoiding open fire. A new saggar was devised by the potter who used a set of cushions. This enabled pottery of various shapes—whether bowl or plate to be fired together. The space of the clay box was fully utilised. It reduced cost and enhanced output. Another method popularly used in the period was finding the temperature of kiln or boiler to guarantee the rate of finished products.

In the Yuan Dynasty underglaze porcelain was fully developed. Blue and white porcelain and underglaze red occupied an important position in pottery manufacture with their excellent colours and glossiness. By the Ming and Qing dynasties pottery technology developed still further. Jingdezhen became the centre of ceramics industry in China, where many civilian and government kilns were situated. The production process became streamlined into a complete and integrated set. The bamboo knife was replaced by the revolving knife (made of iron) in making paste, which raised efficiency. Glazing by sufflation replaced the old method of glazing by immersion. Single glazed colour, such as sacrificial red, ruby red, peacock green, turquoise, bright yellow and sweet white were produced as new colours. The technology of blue and white porcelain and underglaze red consummated, becoming the mainstream in pottery industry. Polychrome glaze took the form of strongly contrasting colours (dou cai) and five-colours (wu cai). Soft-coloured glaze in cloisonne was successfully made during the later phase of this period. The style turned in the direction of over elaborateness. Government kilns paid attention to the choice of better quality clays and filtering. To kaolin were added other materials. The porcelain gained more translucence, smoothness, and glossiness, after firing.

As the inventor and main producer of ceramics, China has long earned the reputation of being the porcelain nation. Ever since the Han and Tang dynasties porcelain has been exported from China to many countries in an endless stream. Porcelain manufacturing method has spread from China to many countries, exerting an important impact on the life of different peoples, as ceramics has both utility value as items for practical use and high artistic value as visual art.

7-1*

7-1

7-2

7-1 New Stone Age red pot, thin-necked, fish-bird design, 21.6cm high, 2.1cm in diameter. The diameter of bottom part is 8.5cm. Unearthed from Beishouling, Baoji, Shaanxi in 1958, it is kept in the National Museum of Chinese History.

7-2 New Stone Age red pottery jar in whorl design, 50cm high, 18.4cm in diameter. The bottom is 15.9cm in diameter. It has a wave and water flow pattern in black and brown. Obtained from Erping, Yongjing, Gansu in 1956, it is kept in the National Museum of Chinese History.

7-3

7-4

7-5

7-5 Shang Dynasty pottery mould used to impress cloud design, after the body of pottery had been made. Measuring 11.7cmX5cm X1.5cm, it was unearthed from Minggong Lu, Zhengzhou, Henan and is kept in the National Museum of Chinese History.

7-3 Neolithic vessel for mixing two kinds of colours, 14.8cm in diameter, 6.5cm high. The bottom has a diameter of 9.8cm. Unearthed from Baidaogouping, Lanzhou, Gansu Province in 1953, it is kept in the National Museum of Chinese History. Spherical in shape, it is segmented in the middle. It is used in mixing colours during pottery manufacture.

7-4 Pottery cushion, 8.3cm X 5.5cm of the New Stone Age. Unearthed from Miaodigou, Shaanxian, Henan in 1956, it is kept in the National Museum of Chinese History.

7-7

7-6

7-6 Schematic diagram showing the clay strip forming technique on a potter wheel. The surface is smoothed even. This is the earliest coil method of the New Stone Age. Using disc of clay, potter rolls out strips of clay, pinching and smoothing. It is used even to this day in ethnic minority areas.

7-7 Relic of pointed bottom of bottle, 9.8cm high and 7cm in diameter (New Stone Age). Unearthed from Shaanxi, it is kept in the National Museum of Chinese History. The red pottery bottom reveals that the bottle was made with clay strip method.

7-8 Schematic diagram showing rotating wheel method. It was adopted in the New Stone Age. Potter formed a cylinder with clay and shaped it. The wheel in early versions rotated slowly, pushed by the hand. The wheel rotated itself due to inertia force. The hands of the potter worked on the clay to make desired pottery. Traces of parallel lines are left on the pottery on both surface and inside. The bottom of the vessel is left with eccentric patterns.

7-9 Traces on black pottery from rotating wheel. The vessel was unearthed from Sanlihe, Jiaoxian, Shandong in 1975. It is 7.8cm high, 22.7cm in diameter. Bottom 16.4cm in diameter. Belonging to the New Stone Age, it is kept in the National Museum of Chinese History.

7-8

7-9

7-9*

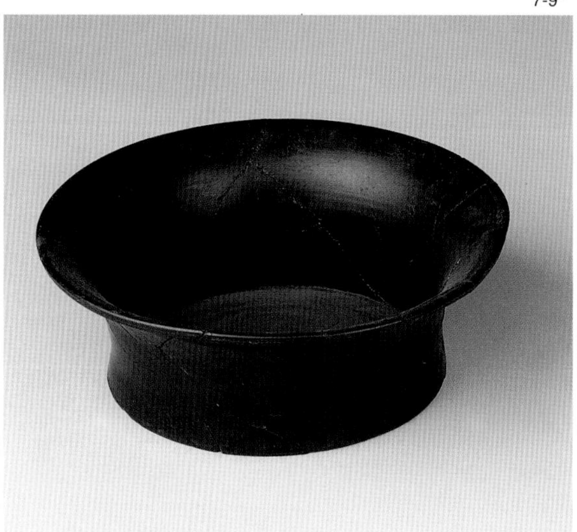

7-10

7-10 New Stone Age reticulated black pottery cup. It stands 19.5cm high and was unearthed from Anqiu, Shandong in 1957. This kind of ware could be as thin as 0.2mm. After firing,the body of pottery was subjected to a process of smoking with carbon grain by means of quick rotating wheel, making it black. The surface of the pottery was burnished and smoothed till it shone. It is kept in the National Museum of Chinese History.

7-11 New Stone Age white pottery gui (wine vessel) with cord pattern. It belongs to late New Stone Age. It measures 33.5cm high with a diameter of 12cm. The mouth is 11cm. Unearthed from Rizhao, Shandong in 1955, it is kept in the National Museum of Chinese History. The 1.5-1.72% iron which it contains makes it a white porcelain. It was fired in kiln at a temperature of 1,000 degrees Celsius. Its chemical makeup is close to porcelain clay and kaolin. China is the first country to use kaolin to make pottery ware.

7-11

7-12

7-13

7-12 Grey pottery he(wine container), with a belly 14.2cm in diameter and stands 20.2cm high. The diameter of the largest leg is 14.1cm. Unearthed from Erlitou, Yanshi, Henan in 1972-73, it belongs to the Xia Dynasty (c.21st-16th century BC).

7-13 Shang Dynasty hard pottery dou (bowl), with impressed design, 18cm in diameter and 25cm high. The belly is 27.3cm in diameter. Unearthed from Huangtucang, Fujian in 1975, it is kept in the National Museum of Chinese History. This ware was discovered in both the Yellow and Yangtze River Valleys. The earliest ware can be traced to the Xia Dynasty. Hard pottery is finer and harder than ordinary porcelain or sandy ware, requiring higher temperature for firing. The designs are printed by means of pottery mould. The surface contains printed patterns. Hence the Chinese term yin wen ying tao, meaning hard pottery with impressed pattern.

7-14

7-15

7-16

7-14 Thin proto-porcelain dou, unearthed from Tawan, Luoyang, Henan Province, is kept in the National Museum of Chinese History. The dou is 15.4cm in diameter and 7.9cm high. The leg is 7.8cm in diameter. As iron is found in its glaze, it becomes green after firing. Proto-porcelain made its appearance in the Shang and Zhou dynasties. Proto-porcelain green ware was produced mainly in the middle and lower reaches of the Yellow and Yangtze rivers. It was developed on the basis of white pottery and hard pottery with impressed designs. The ones made in Zhejiang and Jiangsu have fine body with even walls, regarded as superb wares.

7-15 Unearthed from Zhongzhou Lu, Luoyang, Henan this green porcelain jar has four rings, 15.4cm in diameter, 16.8cm high with leg 7.8cm in diameter. With a swelling body and straight mouth, it is the earliest consummated ceramics. It has scientific value in the study of the history of ceramics invention.

7-16 This celadon lotus zun (wine vessel), 61.5cm high, 16.4cm in diameter, was unearthed from a grave, whose occupant was named Feng, in Jingxian,

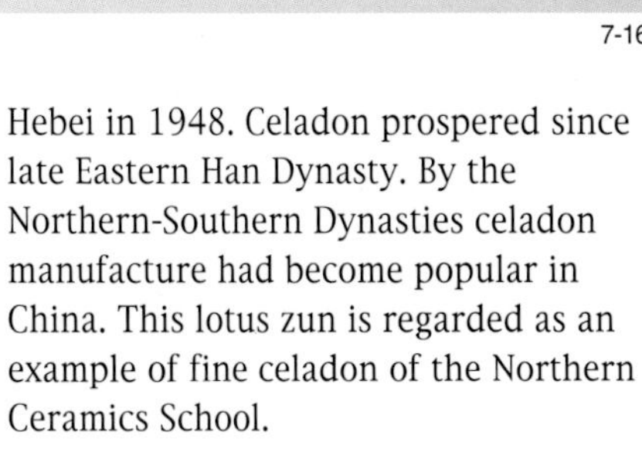

Hebei in 1948. Celadon prospered since late Eastern Han Dynasty. By the Northern-Southern Dynasties celadon manufacture had become popular in China. This lotus zun is regarded as an example of fine celadon of the Northern Ceramics School.

7-17

7-17 Tang Dynasty porcelain in normal white, 22.2cm high, 6.9cm in diameter and diameter of foot 7.3cm. Unearthed from Shaanxian, Henan in 1956-57, it is kept in the National Museum of Chinese History. Attaining consummation in the Tang Dynasty, white porcelain requires low content of iron—not exceeding 1%. Its paste is fine and thinly glazed. It is either white or green.

7-18 This olive green bowl, 6.1cm high and 23.8cm in diameter, was unearthed in Fa Men Si, Fufeng, Shaanxi and is kept in the National Museum of Chinese History. It is an excellent specimen among green ware made in Yue Kiln, which produced mainly green, blue and green yellow glaze plus gold and silver adornments.

7-19 This blue and white porcelain, 20cmX10.5cm was unearthed from Yangzhou, Jiangsu in 1983 and is kept in the Yangzhou City Museum. Designs are drawn on the paste containing cobalt. A translucent glaze is added. Cobalt becomes blue after heating in high temperature.

7-20 Three-coloured woman figurine, 45.3cm high, unearthed from Xi'an, Shaanxi 1957. The Tang Dynasty ware is low temperature glaze. Due to the presence of lead the temperature for baking is reduced. The glaze spread all over the body during firing. Due to lead the colours mix. They are brown, black, white, yellow and green which form a motley of colour schemes. The three colours are yellow, green and brown.

7-18

7-19

7-20

7-22

7-21 Tang Dynasty three-coloured tomb animal guardians, 52.5cm X17.5cm and 53.5cmX17cm respectively. It was unearthed from Hansenzhai, Xian, Shaanxi in 1955.

7-22 Jun Kiln flower vase is purple rose mixed with red in colour, begonia style, 14.4cm high. The diameter is 24.6cm. The width of the mouth is 19.5cm. The leg is 13.3cm long and 10.5cm wide. It is kept in the National Museum of Chinese History. As the glaze contains oxidized copper, it undergoes furnace transmutation so that the blue becomes red or purple. The colour is very bright.

7-23 Jun Kiln sky blue glazed saggar bowl, rose pattern, with a diameter of 25cm. It is placed within a saggar which is 11.8cm high with a diameter of 29.8cm. It is kept in the National Museum of Chinese History. The saggar is an important tool, which makes clay heated in an even manner. The colour does not change by smoke. More paste can be packed into the kiln by means of the saggar.

7-23

·21

7-24

7-25

7-25*

7-25 Yaozhou mould for impressing chrysanthemum design, 5cm high and 16cm in diameter. Kept in the National Museum of Chinese History, it was used to make paste. The design was impressed on the inner walls of bowls.

7-26 Song Dynasty saggar for grinding glaze material, 24.2cm in diameter and 14cm high. The leg is 9.3cm in diameter. Obtained from the ancient kiln in Longquan, Zhejiang, it is kept in the National Museum of Chinese History. Its inside has grooves with fishing net design, used to grind glaze materials.

7-26

7-28

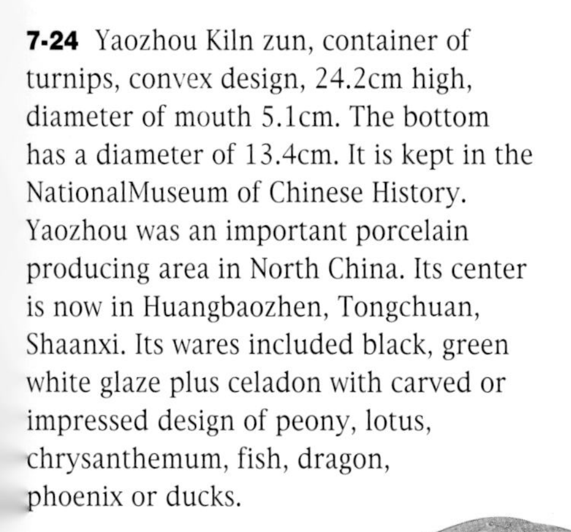

7-24 Yaozhou Kiln zun, container of turnips, convex design, 24.2cm high, diameter of mouth 5.1cm. The bottom has a diameter of 13.4cm. It is kept in the NationalMuseum of Chinese History. Yaozhou was an important porcelain producing area in North China. Its center is now in Huangbaozhen, Tongchuan, Shaanxi. Its wares included black, green white glaze plus celadon with carved or impressed design of peony, lotus, chrysanthemum, fish, dragon, phoenix or ducks.

7-27

7-27 Arched cutting tool, 6.9cmX0.4cm, obtained from ancient kiln in Longquan, Zhejiang. It is kept in the National Museum of Chines History.

7-28 Glaze top tile, 5.2cm high and 7.1cm in diameter. The diameter inside is 5cm and the depth is 4.5cm. Obtained from ancient kiln in Longquan, Zhejiang, it is kept in the National Museum of Chinese History. Like an axle, it is a device which supports the revolving wheel.

7-29

7-30

7-29 Longquan Kiln soft-green stitched-band incised lotus bowl, 5.8cm high, with flared mouth 14.3cm in diameter and leg 3cm in diameter. It is kept in the National Museum of Chinese History. Longquan porcelains, either incised or applied design, were exported by sea to many countries in the Yuan-Song period. Longquan was one of five leading ceramics centres.

7-30 Ming Dynasty Longquan ware, convex peony-plum bottle, 38cm high and 4.5cm in diameter. Bottom is 10.5cm in diameter. It is kept in the National Museum of Chinese History.

7-31 Yuan Dynasty blue and white porcelain, dragon/cloud design, pear-shaped. It is kept in the National Museum of Chinese History. With underglaze colours, the design is drawn in glaze containing cobalt, to which is added transparent glaze. After firine in high temperature, the cobalt becomes blue. Lots of ware have been found under blue and white porcelain—bottle, plate, bowl and pot. The manufacture of the ware began with the Tang, carried on in the Song, and thrived in the Yuan Dynasty.

7-31

7-32

7-33

7-32 Illustration on porcelain manufacture in Jingdezhen. There are six processes: moulding, pasting, drawing on paste, glazing, packing clay for baking and painting on pottery. The book *Jingdezhen Tao Lu* (Catalogue of Jingdezhen Ceramics) gives an account of porcelain manufacture in the Qing and Ming dynasties.

7-33 Lang Kiln zun in red, resembling gold and precious stone, 42.8cm high and 11.8cm in diameter. The leg has a diameter of 14.6cm. It is kept in the Museum of Chinese History. The red deepens in the lower part. There are cracked patterns. The Lang Kiln red zun is a representative work of Qing Dynasty porcelain in the reign of Kangxi.

7-34 Qing Dynasty vault-of-heaven vase, posy design, 48.8cm high and 12cm in diameter. The bottom is 11.8cm in diameter. Kept in the National Museum of Chinese History, it is underglaze blue and white porcelain combined with overglaze paintings. The painting material contains iron, copper and cobalt. The finished products are gorgeous and bright. The drawings are wide ranging in content.

7-34

7-35

7-36

7-35 Famille rose zun in deer design, 44.6cm high, 16cm in diameter, and diameter of leg 24.3cm. Famille rose is a kind of overglaze colour with white glass powder as pigment. A wide variety of famille rose was produced in the reign of Qianlong, Qing Dynasty, which witnessed development in technology in famille rose or famille verte.

7-36 Gourd-shaped bottle, underglaze red, with design symbolic of good fortune and long life, 31.2cm high, diameter of mouth 5.1cm and diameter of leg 9cm. made in the reign of Qianlong. It is kept in the National Museum of Chinese History. Patterns are drawn on paste with glaze containing copper, to which is added transparent glaze. It is baked once in high temperature. The ware began with Yuan and developed in the reigns of Kangxi and Yongzheng, Qing Dynasty.

7-37 Hexagonal blue and white porcelain vase, 68.5cm high and 20.3cm in diameter. The leg has a diameter of 23cm. It is kept in the National Museum of Chinese History. Pigment with cobalt from Zhejiang was mainly used to make blue and white porcelain in the reign of Qianlong. The porcelain is bold and bright in colour. The one in deep colour is blue and somewhat black.

7-37

7-38

7-38*

7-38 Club-shaped vase in colours, 48.8cm high and 12cm in diameter. The bottom is 11.8cm in diameter. Decorative porcelains are of two kinds: overglaze colours and underglaze blue and white. This vase is overglaze. On baked white glaze pattern is drawn, requiring a second firing at 700-800 degrees Celsius. The ware has different colours--red, green, yellow, blue or purple. Only two or three colours are used on one vase. This vase containing figures, made in the Qing Dynasty, is kept in the National Museum of Chinese History.

7-39

7-39 Purple earthen pot, 10cm X19.3cm X 12cm, is kept in the National Museum of Chinese History. It is made of purple sandy earth with a relatively high proportion of iron content. The purple sandy earth is found in Yixing, Jiangsu. Sometimes it is purple red, light yellow or purple black. Of fine and tough quality, it can be easily pinched into desired shape. It does not have to be packed into a saggar and can be baked directly over fire at a temperature of 1,000 degrees C. The original colour is dark, simple and heavy. No glaze is required for this kind of ware. Purple earthen vessels made in the Qing Dynasty, of highest quality, were presented to the emperor as gift.

Map of Distribution of Famous Ancient Chinese Kilns

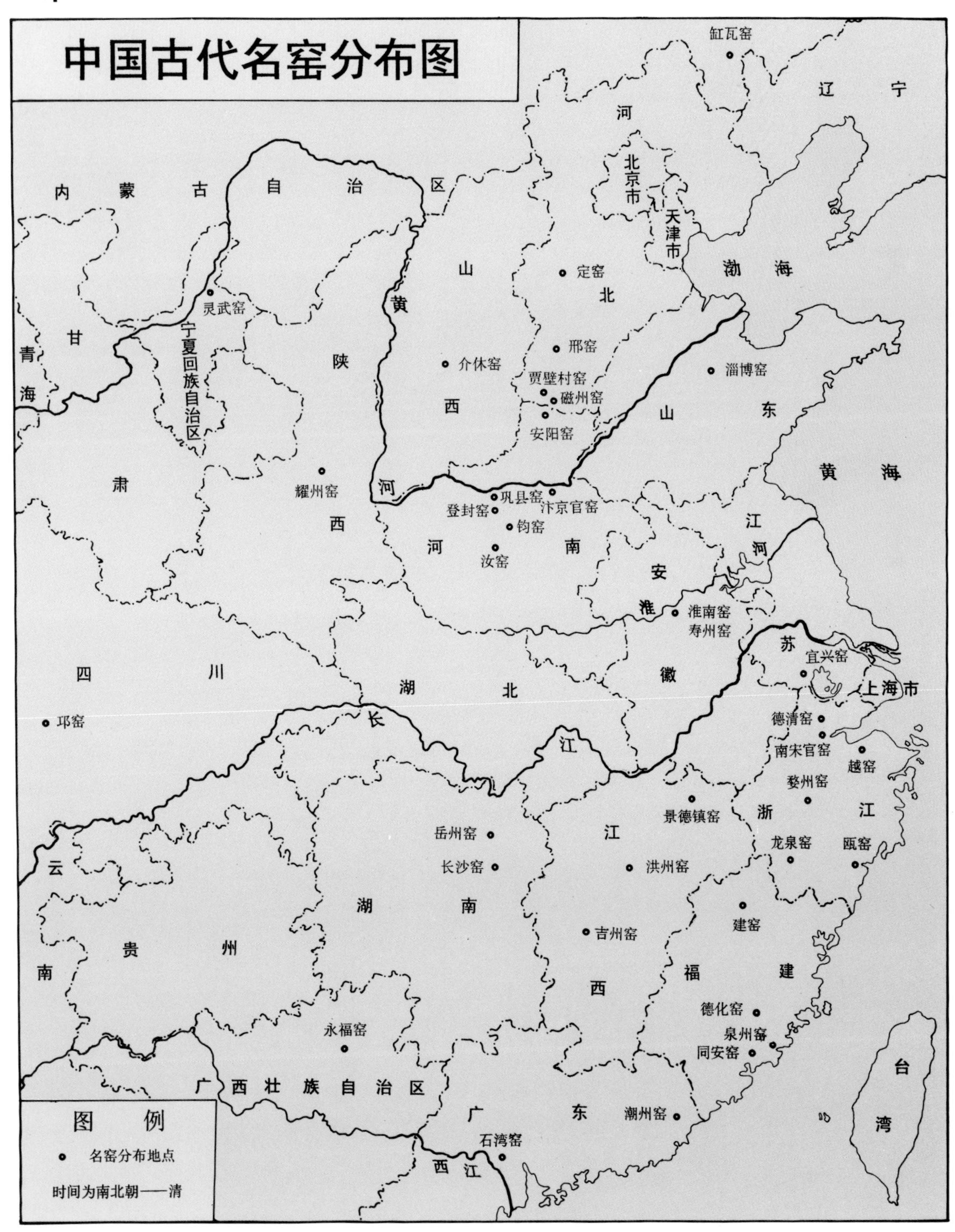

TEXTILES

Ancient Chinese textiles can be traced to the middle of the Palaeolithic Age. The primitive ancestors made simple ropes and fishing nets required for hunting and gathering, which were their main pursuits. Sewing technique was mastered at a later stage. Threads were woven into various types of fabrics. Entering the Neolithic Age the Chinese made simple spinning machines as well as looms (yaoji, literally waist machine). The loom was a disc like device, in the centre of which was a plant used as a stick.

By means of coupling a woman spun the fibre into yarn. She sat on the ground and used the foot to pedal the treadles to interlace lengthwise threads and crosswise ones. The warp was picked through the opening of the shuttle. The crosswise threads were beaten back and ran through to bring threads to uniform level. The interlacing went on. People gained some knowledge of natural fibre during this time. Relics of hemp woven piece at the bottom of porcelain ware unearthed in the Neolithic ruins of Banpo, Xi'an, bear strong evidence to this fact.

China learned how to handle sericulture in the late phase of the New Stone Age. Silkworm cocoons cracked open by man have been discovered in the ruins of the Yangshao Culture in Xia County, Shaanxi Province. Silk pieces and silk bands have been found in the ruins of the same period in Wuxing, Zhejiang Province. Sericulture made considerable progress as man entered class society. Inscriptions on tortoise shells and oracle bones of the Shang Dynasty reveal the words jian (cocoon), si (silk) and bo (silk). Jade pieces in cocoon shape have been excavated from Shang Dynasty graves in Anyang, Henan Province and Yidu, Shandong Province. In the Western Zhou Dynasty the raising of silk worms became a popular pursuit in the Yellow River Valley.

The Book of Poetry had something to say about sericulture and silk fabric. It mentions a rather big mulberry field. Unearthed bronzewares of this period contain designs of women collecting mulberry leaves. Simple traditional reeling machines, spinning wheels and looms appeared. Foot-pedalled looms emerged in the Spring and Autumn Period at the latest. Humans were the source of generating power then. They stretched threads on the weft while pedalling the loom. Following the expansion of textile production those in the trade became more and more professional. Spinning, weaving and dyeing were done separately by different people. Having bolstered the artistic value, the products became commodities. From the ruins of the Yin Dynasty, Anyang, Henan we see traces of thin silk designs and fretwork. Jacquard technique was developed, which was a contribution by China to textile industry. Silk fabric greatly increased in output. According to *The Book of Poetry*, silk gauze, thin silk, damask and brocade had come into existence in the Shang and Zhou dynasties. Natural plants were used as dye stuff to dye fabric in single or different colours. The hemp excavated from a grave in Jingyang, Shaanxi shows fine texture. Geometrical designs in red, white, black and yellow are seen on a painted cotton curtain excavated from a Zhou Dynasty grave in Luoyang, Henan, reflecting the level of textile technology of the era.

In the wake of the Qin and Han dynasties silk and textile technology developed effectively. The foot-pedalled loom, with a loop in the middle to receive warp thread for making twilled basket weave, became popular, followed by jacquard machine with a multiplicity of heald twines and foot pedals. The Han Dynasty not only produced silk as thin as a cicada's wing, but also refined fabrics, such as damask and gauze. The brocade shows wide-ranging variety of complicated designs. A square meter of plain silk, weighing only 15 gm, excavated from the Mawangdui No.1 grave, Changsha, Hunan Province reflects the advanced level of technology. In the same grave has been found brocade pasted with down, made by heddles for guiding warps with a double lengthwise jacquard machine. The more coarse warps become circles. The fret is raised above the fabric. This gives a marked three-dimensional effect. The Eastern Han Dynasty jacquard, operated by two persons, had 120 heald twines. One person 3 ft above, made designs according to drawing on a machine. The weaver below worked on the weft in close co-ordination with him.

In the Tang and Song dynasties (618-1279) Chinese textile industry consummated in skill day by day. The jacquard was finalised in design. A big machine drawn by the painter Lou Shou of the Southern Song Dynasty depicts the machine, operated by two persons with double warp axle and ten heald twines. One man works on top and the other below. The quality of silk fabric was high, with a wide ranging variety of products. Some Chinese silk designs absorbed Central Asian style with fowls and animals as subject matter, as are seen on brocade with fowls and animals forming a connected picture and granulated phoenix with crown, unearthed in Turpan, Xinjiang. Other fabrics show a series of pictures depicting Chinese life in 12 months, and lantern brocade with gold thread as weft. A new technology adopted in silk weaving is tapestry of cut silk, with design made on warp and details woven in colour by means of small shuttle. Described as connected warp and broken weft, this method produces hundreds of colours and beautiful patterns on silk. It is often used to copy paintings and calligraphy on silk. The outstanding works of the Southern Song Dynasty are ducks in a lotus pond by Zhu Kerou and plum and magpie by Shen Zifan.

The book, *Zi Ren Yi Zhi* (Heritage in Manufacture of Hometown People) by Xue Jingshi, a carpenter of the Yuan Dynasty, describes the structure and manufacture of loom, jacquard, and gauze loom. Size and sample drawings are given. The book has great historic value. A contemporary, Huang Daopo, studied the textile technique of the Li ethnic group on Hainan Island and improved the cotton gin, flicking and fluffy tool and cotton- spinning device. She gave great impetus to textile industry in the lower Yangtze River Valley. Wang Zhen devoted much space to describing silk, cotton and hemp implements in his *Nong Shu.* Chinese textile industry made more technical advances in the Ming and Qing dynasties as silk was produced in great quantity in Suzhou, Hangzhou, Chengdu, Guangzhou and Fujian. The silk was consumed at home or exported abroad. There were a number of government-operated mills to make silk for royal household use. Civilian mills thrived. The jacquard technique improved, with the replacement of five-layer warps by four- layer ones. Silk became thinner, catering to more practical use. Less raw materials were consumed. Varieties improved and new styles appeared day by day. There were 17 kinds of embroidered works on dressing table for ladies alone. The technique heralded the more advanced jacquard design later. The book *Tian Gong Kai Wu* by Song Yingxing summed up various types of textile machines. The double-sided velvet unearthed in the Ming Tombs, Beijing indicates 64 warps and 36 wefts per sq. cm. of fabric—very closely woven indeed. Silk industry thrived in Suzhou and Hangzhou in the Qing Dynasty. As many as 10,000 households in Suzhou engaged in silk weaving or embroidery while the sound of the loom was heard in every household in Hangzhou. The silk industry spread to Nanjing, Guangzhou and Foshan. In Nanjing over 30,000 looms were operating to turn out satin. As a machine the loom was made up of 130 component parts.

In a word, China is the first country to embark on sericulture and silk industry. She is credited with inventing or laying the foundation for many machines—silk reel, twill weaving looms, foot-pedalled heddles for guiding warps, and jacquard. Silk was exported from China in the Han-Tang period by overland route, known as the Silk Road. After the Tang Dynasty, China shipped silk abroad via the maritime Silk Road from Guangzhou, Quanzhou and a host of other ports as a result of the development of shipping and navigation.

8-1

8-2

8-3

8-1 Revolving pottery wheel belonging to the New Stone Age. It stands 1.7cm high and has a diameter of 6cm. Unearthed from Banpo Village, Xi'an, Shaanxi, it is kept in the National Museum of Chinese History. The main part of the spindle, the wheel, uses its weight and inertia force to twist hemp, silk and wool. It spins cotton yarn (rough or fine) as well.

8-2 New Stone Age pottery wheel, 3.5cm in diameter and 2.1cm high, unearthed in Hemudu, Yuyao, Zhejiang in 1978.

8-3 Using the pottery spinning wheel—a schematic diagram.

8-4 Yaoji—the waist machine in operation—an illustration.

8-5 This New Stone Age pottery saggar is 7.2cm high, 14.5cm in diameter and the bottom has a diameter of 6.3cm. Unearthed from Banpo Village, Xi'an, Shaanxi, it is kept in the National Museum of Chinese History. The texture on the saggar is in plain weave. The warps and wefts are arranged evenly. The yarns are rather rough, which tells us that the saggar makes cushions which are used for the manufacture of pottery.

8-4

8-5

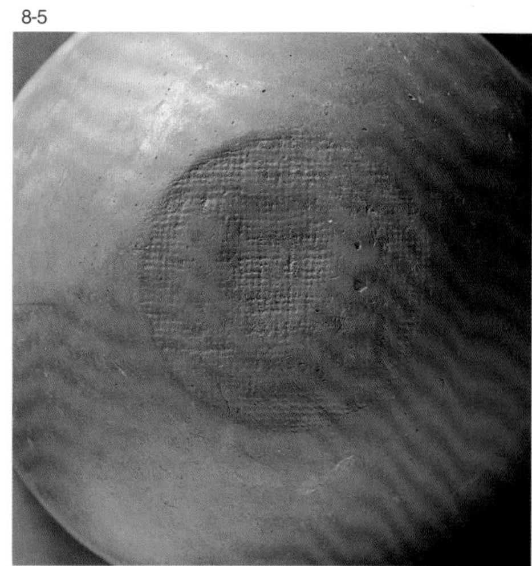

8-5*

8-6* 8-6

8-7

8-6 Traces of silk fabric on Shang Dynasty copper piece. It is 19.8cmX14.1cmX0.1cm. This piece was unearthed frow Dasikong Village Anyang, Henan in 1953 and in kept is the National Museum of Chinese History.

8-7 Replica of a white silk, closely woven with 44 warps and 18.5 wefts per cm of fabric. It is part of a silk piece on Shang Dynasty copper piece.

8-8

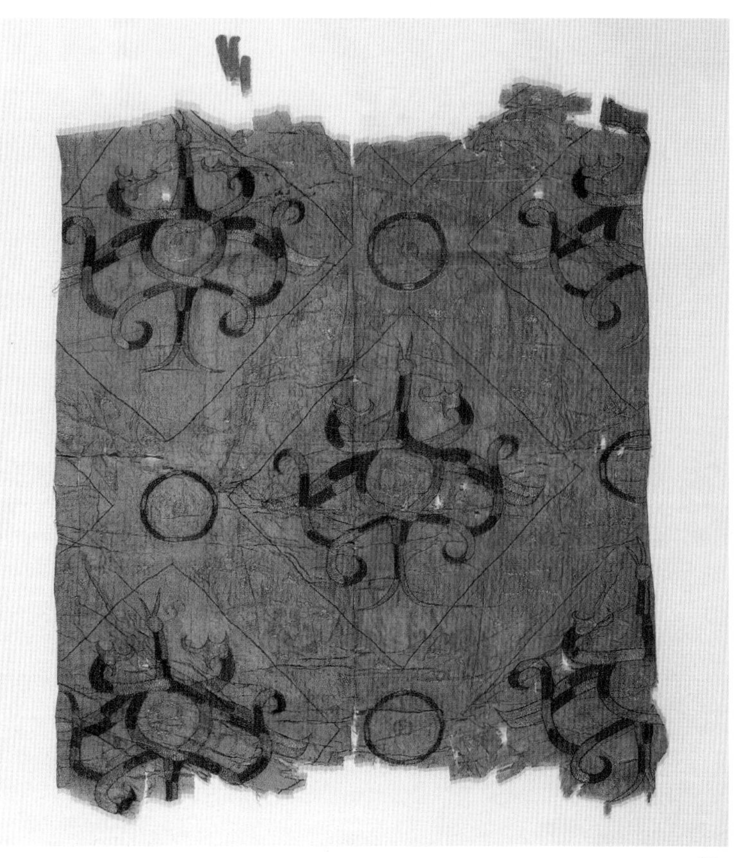
8-9

8-8 Tao (silk rope or band), with dragon and phoenix design, 5.7cmX0.42mm. Unearthed from Mashan No. 1 Chu Grave, Jiangling, Hubei, it is kept in the Jingzhou Prefectural Museum, Hubei. The tao was used to make collar on Chinese dress. On the weft is design against a background of plain weave. The dragon and phoenix design is arranged in upper and lower parts of the fabric symmetrically. The tao belongs to the mid-Warring States Period.

8-9 This mid-Warring States Period brocade is 49-50cm wide. The edge is 0.6cm. It is 0.33mm thick. Unearthed from Mashan No. 1 Chu grave Jiangling, Hubei is in kept in the Jingzho Prefectural Museum, Hubei. It is jacquard on warp threads. Twelve complete patterns repeat themselves carried by 12 bands on warps. Each band has a different pattern. The warp threads are in two colours. Altogether four colours appear on warps as the pattern re-emerges. Arranged from left to right geometrically they contain phoenix, bird and wild duck pattern, in two colours.

8-10

8-10 Damask with confronting birds design, Western Han Dynasty, 40cmX19.5cm. Unearthed from Mawangdui No. 1 Han Dynasty grave, Changsha, Hunan in 1972, it is kept in the National Museum of Chinese History. The damask is woven first and dyed later. The patterns are woven with threads in vermilion, brown, deep green, deep blue and yellow, using queue strand stitch. They are clouds looking like ears of grain and coiled grass. The complete piece of damask is made up of designs in five and half units. Each unit measures 9.5cm long and 7.5cm wide.

8-11

8-11 Northern Dynasties brocade with lantern and tree designs, 20cmX17.1cm. Unearthed from the No.303 Grave in North District, Astana, Turpan, Xinjiang in 1959, it is kept in the Xinjiang Uygur Autonomous Region Museum. It has double textured warps, separated by white-blue and white-green backgrounds. On orderly arranged coloured bands are trees arranged horizontally at equal distance from each other.

8-12 This is a brocade with animal design set in squares, 18cmX13.5cm. Unearthed from the No.99 Grave, North District, Astana, Xinjiang in 1968, it is kept in the Xinjiang Uygur Autonomous Region Museum. Double textured warps contain threads in red, yellow, blue, white and green with designs of lion, ox and elephant. It belongs to the Northern Dynasties (420AD-581AD).

8-12

8-13

8-14

8-13 Tang Dynasty (618-907) 9-point green silk gauze, 18cmX5cm, was unearthed from the No. 1 Tibetan Grave, Dulan, Qinghai in 1983. It is kept in the Qinghai Cultural Relics and Archaeology Institute. The traditional tie-and-dye method on four warps is used to weave the fabric.

8-14 Tang Dynasty cut silk in ten designs against blue background, 27cmX8.5cm, was unearthed from the No. 1 Tibetan Grave, Dulang, Qinghai in 1983. It is kept in the Qinghai Provincial Cultural Relics and Archaeology Institute. It is woven in traditional way, known as "continue with warps while break the wefts". The designs are woven in wefts, yellow, green, blue, white and brown. In the area with the same colour the silk is cut in breaks, showing reticulation effect.

8-15 Upholster in cut silk with longevity and good fortune design, 133.5cmX52cm. It is kept in the National Museum of Chinese History. The upholster has a blue border and a red background. There are three sets of designs, showing longevity bottle and floral patterns.

8-15

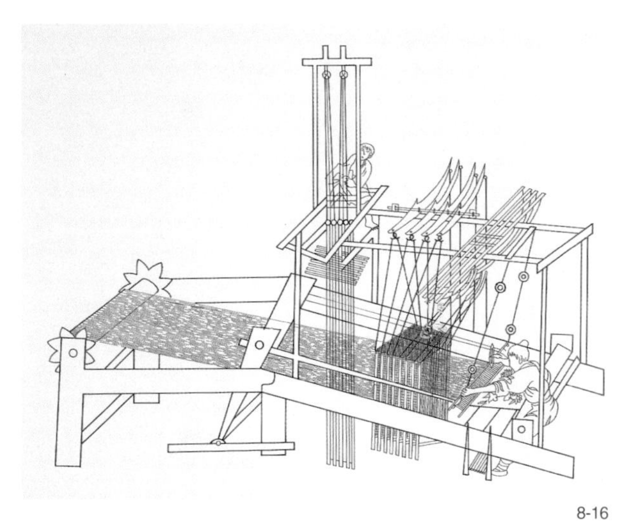

8-16

8-17

8-16 Diagram showing Ming Dynasty (1368-1644) loom. The diagram is based on information from the book: *Tian Gong Kai Wu* by Song Yingxing. The famous Sichuan brocade and Nanjing brocade with cloud designs are made with this kind of loom.

8-17 Zhangzhou satin, 141cmX58cm, is kept in the National Museum of Chinese History. An outstanding product of the

8-18

Qing Dynasty in Dicentra Spectabilis (peony) design with pale mauve coloured background, it is evolved from twill weave. Smooth and glossy, the satin is good material to make bedroll cover or dress. It earns its name from its place of production—Zhangzhou (Fujian Province).

8-18 Jiangling (Nanjing) silk spinning wheel of the Ming-Qing era, 30cmX105cmX44cm, kept in the National Museum of Chinese History. Its structure is made up of wheel (diameter 2 meters) and a rack, 2.5 meters long, on which are placed 56 spindles. It is operated by man. Silk threads are spun into yarn. The spinning wheel is probably evolved from the big spindle of the Yuan Dynasty.

8-19 Blue-white cotton with impressed design, 48cm X89cm, unearthed from Minfeng, Xinjiang in 1959. It is kept in the Xinjiang Uygur Autonomous Region Museum. It has a plain weave, 18 to the warp and 13 to the weft per each cm of fabric.

8-19*

8-19

8-20

8-20 Figurine of the Northern Dynasties (420AD-581AD), 20.5cm high. Unearthed from North District, Astana, Xinjiang in 1964, it is kept in the Xinjiang Uygur Autonomous Region Museum. When found, it wore cotton jacket and trousers. Most figurines unearthed in the region are dressed in cotton jackets and trousers, since cotton fabric was rather popular in the region throughout the Northern-Southern Dynasties.

8-21 Northern Dynasties (420AD-581AD) cotton piece, with impressed blue tortoise shell design, white background, measuring 16cm long and 10.5cm wide. Designs are carved on wood, and printed on the fabric with a clamping device. Unearthed from Yutian, Xinjiang in 1959, it is kept in the Xinjiang Uygur Autonomous Region Museum.

8-22 Tang Dynasty hemp, 198cmX43cm, unearthed in Turpan, Xinjiang in 1967. It is kept in the National Museum of Chinese History. The texture on the warp and weft is plain weave.

8-23 Ming Dynasty Songjiang cotton cloth, 230cmX57.5cm, unearthed from the grave of Du Shiquan, Fengxian, Jiangsu and is kept in the National Museum of Chinese History. Songjiang was a textile center during the Ming Dynasty. Songjiang cotton fabric is of top quality and has a high output, enjoying reputation nationwide.

8-21

8-22

8-23

COPPER AND IRON SMELTING AND CASTING

The mining, smelting, and casting of mineral ores formed the basis of the development of Chinese bronze and iron culture as well as her splendid metallurgical technology.

Minerals were referred to as jinshi (metal and stone) in ancient China, which has a long mining history. Copper tools have been excavated in the ruined site of human habitation of the late New Stone Age. In the ruined site of a Shang Dynasty copper mine, Ruichang, Jiangxi Province, installations of considerable size and molochite have been discovered. Mining technology attained considerable level in the Shang Dynasty, which was known to have developed copper mines and made progress in copper refining.

Copper mining technique was considerably enhanced in the Western Zhou and Spring and Autumn periods. Sites of copper mines belonging to the era have been discovered in Tonglushan, Daye, Hubei Province and Tongling and Nanling, Anhui Province. About 8 shafts or potholes, one crook hole, a complete boiler support and copper axe, copper ben (adze), wooden hammer, wooden shovel and bamboo basket were excavated from the Hubei site. Many ancient furnaces have been discovered. They are the earliest shaft boilers for copper refining, complete with furnace basc,body and stack. At the lower part and in front of the boiler are holes to emit slag and copper. By the side is a blast nozzle. The structure of the boiler is rather advanced for this early age, we must say.

In Tonglushan, Daye, Hubei Province the site of a copper mine of the Warring States Period was discovered with shaft, crooked hole and ribbing. The technique of mine support was enhanced to give guarantee to safety. Axe, hammer and drill made of iron plus wooden files have been found as have been baskets made of bamboo and creeping plant, wooden winch (for hoistine minerals), water trough and wooden barrels for drainage purpose. These tools tell us that a whole series of complicated problems in mining—providing support to thirling, ventilation, transport, drainage, etc. had been solved by the Chinese as early as the Warring States Period over 2,000 years ago. Casting—the technique of heating ores, was mastered by man in early times.

Bronze ware was made in the Xia Dynasty, as testified to by the excavation of a stone mould for casting bronze in a Xia site in Dongxiafeng, Xiaxian, Shanxi Province and of a Xia copper tripod vessel with handle and open spout from Erlitou, Yanshi, Henan Province. Bronze casting entered its peak in this period. We see a nipple pattern bronze rectangular ding (cooking vessel), a large piece of work by casting, of early Shang (excavated from Zhengzhou, Henan Province). In middle and late Shang Dynasty bronze vessels were manufactured in large numbers, complicated in shape and design and superbly made, including the huge Simuwu ding (cooking vessel), weighing 832.84 kg excavated in the Yin ruins, Anyang, Henan Province. It is the heaviest bronze discovered so far. Other bronze wares include zun (cup) with four-sheep, dragon-tiger, and tiger-eating-man designs—very vivid and lifelike. Shang Dynasty bronze wares are too numerous to mention. They reflect the production capacity and technical level of casting of the era.

Inlaid decoration technique on bronze ware began with the Shang Dynasty, as bronze ge (dagger-axe) and yue (axe), inlaid with gold, silver and emerald have been unearthed. The technique of making copper inlaid with designs had been adopted. It made progress in the Sprine and Autumn Period. Copper wares inlaid with designs have been excavated on many occasions. The copper pot inlaid with design showing a hunting scene was excavated from Jiagezhuang, Tangshan, Hebei Province. The technique of making gold and silver inlaid wares appeared in the Sprine and Autumn Period, the most important example of which is the fou, container with cover and ear-shaped handles for holding wine or water, a relic left behind by Luan Shu, a minister of the State of Jin. About 40 words inlaid in gold are found on the upper part of the container.

Gold coating is a method that covers bronze ware with gold (together with mercury). It was first used in the Warring States Period and became popular in the Western Han Dynasty. The bronze, covered with a gold coating and mercury, allows mercury to evaporate after heating. The gold is retained on the surface of the bronze ware. It is pressed and made smooth. This kind of bronze ware is very splendid and its colour keeps fresh. The zun (wine vessel) with gold-coated bird design is a representative work of this technique during the peak era in the Western Han Dynasty (third reigning year of Yongguang).

Bronze wares of the Shang Dynasty were for the most part made in pottery mould. The object was sculptured. The interior of the mould and the outer shell were made in accordance with the shape of the object. Copper was melted and poured into the mould. The mould for making ritual vessels unearthed in Yin ruins, Anyang, Henan Province has been preserved intact. It reflects the level of mould technology of the Shang Dynasty.

The modern mould now made in China was called lost wax by ancient Chinese, who invented the cire perdue process—an important contribution to world casting technology. The method was used in the Spring and Autumn Period at the latest. The copper jin (rectangular stand supporting wine vessel) unearthed in the Chu State grave in Xichuan, Henan Province is made in lost wax casting method.

The piling up of many moulds to cast copper began in the Spring and Autumn Period. The earliest such moulds were discovered in a copper casting site of the State of Jin in Houma, Shanxi Province. The moulds of the Warring States Period down to the Han Dynasty have been unearthed on various occasions. Some 500 piled up moulds for the manufacture of horse chariots and harnesses and weights for measurement have been excavated from the Eastern Han Dynasty kiln in Wenxian, Henan Province. From this we gain a glimpse into the maturity of multiple-layer mould and the scale of production of the Eastern Han. As multiple-layer mould made many bronze wares in one firing, it raised efficiency many times.

The discovery and application of iron had an important place and played a very crucial role in the history of the development of mankind. It constituted an epoch-making event. As early as the Shang Dynasty, more than 3,000 years ago, Chinese forebears began using native iron. The iron blade of a copper axe unearthed in Pinggu, Beijing is made of meteoric iron. In late Western Zhou Dynasty iron was used to make weapons. The iron sword with copper handle unearthed from the Guo State Grave, Sanmenxia, Henan Province in 1990 has been ascertained to be man-made iron. More iron wares of the Spring and Autumn Period have been discovered in various locations in recent years, including tools, weaponry and ritual vessels. The iron knife of Sanmenxia and the iron sword with gold handle and gold head unearthed from Shaanxian are representative works. The cast iron ding discovered in Changsha, Hunan Province is the earliest cast iron ritual vessel discovered in China so far.

By the Warring States Period iron casting reached its peak. Casting moulds unearthed in Xinglong, Henan Province include those for casting hoe, pickaxe and sickle, showing consummation of skill in making the metal flexible. Upon examination the tools have all been made supple by means of moulds.

Iron smelting and manufacture became the biggest handicraft departments in the Western Han Dynasty. Iron wares of the period spread to all parts of China. Wall paintings carved on stone reveal iron smelting and blast equipment. The pothole is an important installation. Potholes have been discovered in Gongxian, Henan Province and in the Guxing ruins, complete with furnace base, body and stack. The bigger boiler has a diameter of 1.8m. Rectangular boiler, 5-6m high, contains a 4-meter long shaft. It has a volume of 50 cubic meters. The daily iron output is estimated at 1 ton. Among Han Dynasty ruins was found a boiler for torrefying iron ore. The technique of scorching ore in intense heat began at the end of the Western Han. The raw ores, chopped into pieces, were put into a pre-heated boiler together with charcoal. When the raw ores were in a half molten state, they were stirred to increase contact with oxygen. As charcoal was consumed and diminished, the melting point of the ores was increased. The ores now became crusts. After a process of hammering they became wrought iron or steel. The technique was simple and easy to handle, resulting in increased efficiency in production. After repeated hammering and heating, fine steel was produced at very cheap price. Fine swords and knives undergoing hammering and heating 30 to 50 times are known to have been made in the period.

In the Southern-Northern Dynasties a new technique of mixing raw and wrought iron to make steel emerged. Qimu Huaiwen of Northern Qi made the sutiedao (knife) with this method, which was handed down to later generations. In course of time the stack was improved. The Song-Yuan steaming boiler to melt ore was introduced. The opening of the boiler became smaller, which reduced the loss of heat while the firepot was enlarged, which allowed more ore to be heated, preventing hanging. The lower part of the boiler shrank in size, forming into a belly where heat could be concentrated to make it better able to melt the ore. The furnace could melt 10,000-20,000 catties of ore each time. The equipment for blowing air was improved. A wooden fan was used to blow air, which increased air pressure and the volume of air, enhancing the temperature in the furnace. Output was consequently increased. The air-blowing technique was adopted in China earlier than Europe by more than 500 years. In the Ming Dynasty wooden bellows with a piston for producing a current of air to blow up fire in the furnace was used. The piston blower used by China is earlier than Europe by over 100 years.

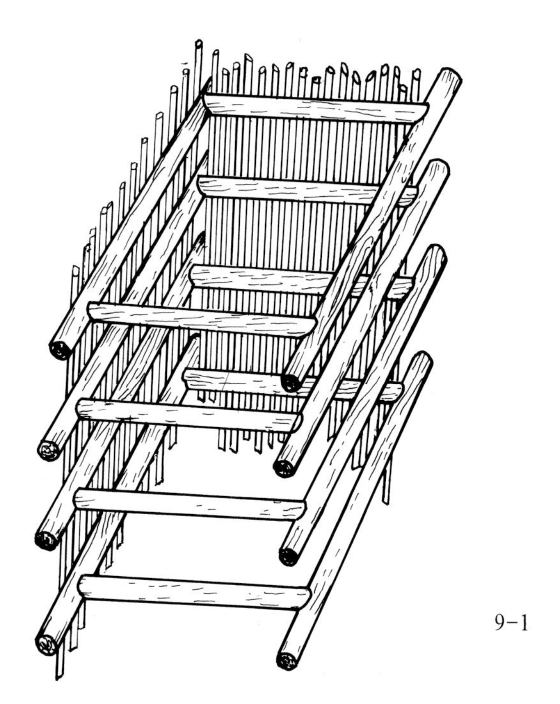

Bowl mouth linked inside (on the same side) support device for furnace.

9–1

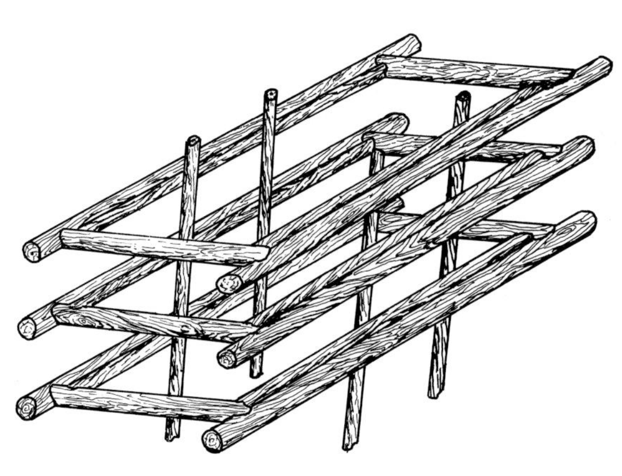

Reinforced bowl mouth linked inside on the same side device for furnace.

9-2

9-1 Schematic diagram on the shaft support device in the site of a copper mine of the Shang Dynasty in Ruichang, Jiangxi.

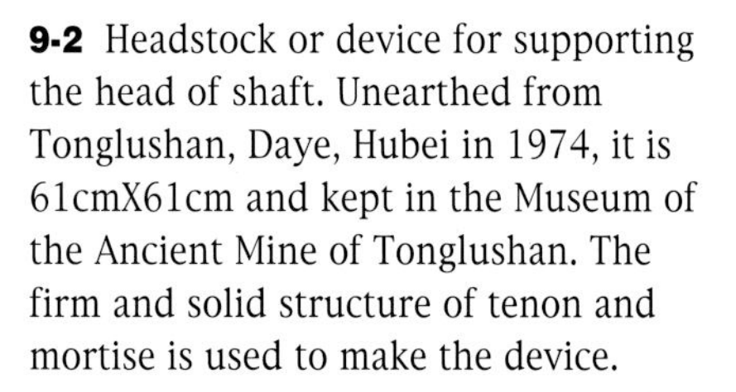

9-2 Headstock or device for supporting the head of shaft. Unearthed from Tonglushan, Daye, Hubei in 1974, it is 61cmX61cm and kept in the Museum of the Ancient Mine of Tonglushan. The firm and solid structure of tenon and mortise is used to make the device.

9-3

9-3 Copper axe, 37cmX32cm, used for mining, unearthed in the site of an ancient mine in Tonglushan, Daye, Hubei in 1974. Belonging to the Warring States Period (475BC-221BC), it is kept in the Museum of the Ancient Mine of Tonglushan.

9-4

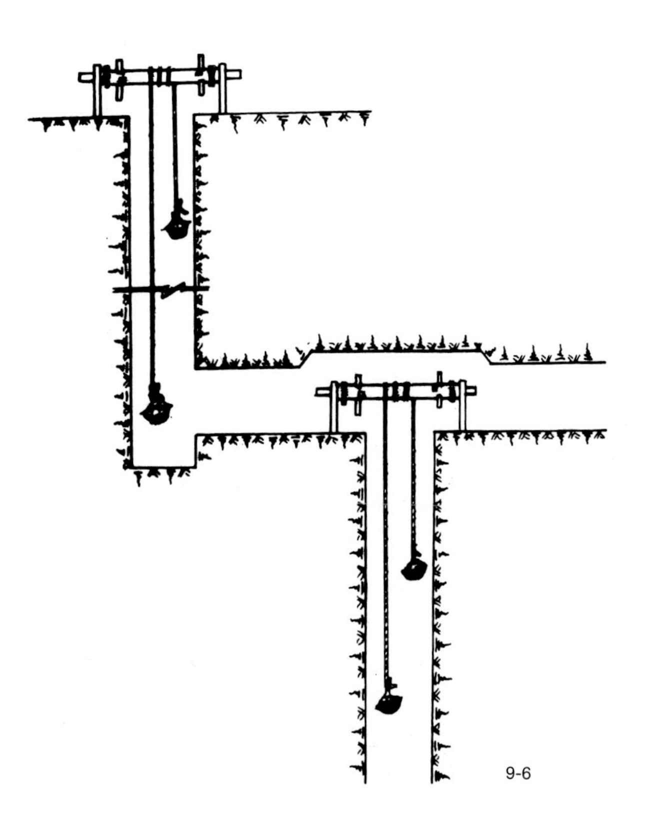

9-6

9-5

9-4 Iron hammer, 15cm high, circumference of belly 9.1cm and 3.5cm in diameter, unearthed in the site of the ancient mine in Tonglushan, Daye, Hubei in 1974. A mining tool, it is kept in the Museum of the Ancient Mine of Tonglushan.

9-5 Wooden barrel, 20.5cm high and 24.5cm in diameter, was unearthed from the site of the ancient mine in Tonglushan, Daye, Hubei in 1974. It is kept in the Museum of the Ancient Mine of Tonglushan. The barrel was used to drain water in the mine.

9-6 Schematic diagram showing silo lifts in the underground chamber in the ancient mine of Tonglushan.

9-7

9-7 Copper ingot, unearthed in the site of an ancient copper mine in Nanling, Anhui. Belonging to the Western Zhou Dynasty (11th century BC-771BC), it is preserved in the Nanling Cultural Relics Management Bureau.

9-8

9-8 Copper knife, unearthed in Lianchen, Yongdeng, Gansu, 3cm long and 1.3cm wide. It is preserved in the National Museum of Chinese History.

9-9 Jue, 12cm high, legs 4.5cm long and bottom 5.5cm wide, unearthed from Erlidou, Yanshi, Henan in 1974. It is preserved in the Archaeology Institute of the Chinese Academy of Social Sciences. It is a bronze wine container. The earliest such container appeared in Majiayao Culture over 3,000 years ago. Complex bronze wares were made in the Xia Dynasty (21st century BC-16th century BC).

9-10 Bronze rectangular ding with nipple pattern, 100cm high and mouth 62.5cm long, unearthed in Duling, Zhangzhai Nanjie, Zhengzhou, Henan in 1974. It is preserved in the National Museum of Chinese History. A giantic bronze ware of the Shang Dynasty, it indicates that casting technique had been greatly improved during the era, with higher temperature and improved clay mould as a prerequiste, which made the manufacture of such an item possible.

9-9

9-10

9-11 Four-sheep zun, 58.3cm high, with length of mouth 52.4cm and width of mouth 52.4cm, unearthed from Ningxiang, Hunan in 1938. It is preserved in the National Museum of Chinese History. A unique and intricate piece of work, it requires superb craftsmanship in making. The fact that it was successfully made shows the high level of technique attained by the Shang Dynasty (16th century BC- 11th century BC). Zun is a wine container.

9-11

9-12

9-13

9-13*

9-12 Tuan fang yi (rectangular casket-shaped vessel with cover surmounted by knob), Western Zhou Dynasty ritual vessel, 22.8cm high, with mouth 14.3cm long and 10.9cm wide unearthedin Meixian, Shaanxi in 1955. It is preserved in the National Museum of Chinese History. It was made by pouring melting ore into many pottery moulds during casting.

9-13 Spring and Autumn Period pot in red metal, inlaid with hunting scene. The pot is 36cm high and its mouth has a diameter of 18.7cm while the diameter of bottom is 14cm. Unearthed in Jiagezhuang, Tangshan, Hebei in 1951, it is preserved in the National Museum of Chinese History. Inlaid work began with the Shang Dynasty and reached maturity in the Spring and Autumn Period. The hunting scene was carved beforehand in the mould. After casting, grooves showing hunting scene were left on the body. Red copper wires were inlaid in grooves, which were beaten to make them even. Rubbing was added. The end result was a nice piece of bronze ware.

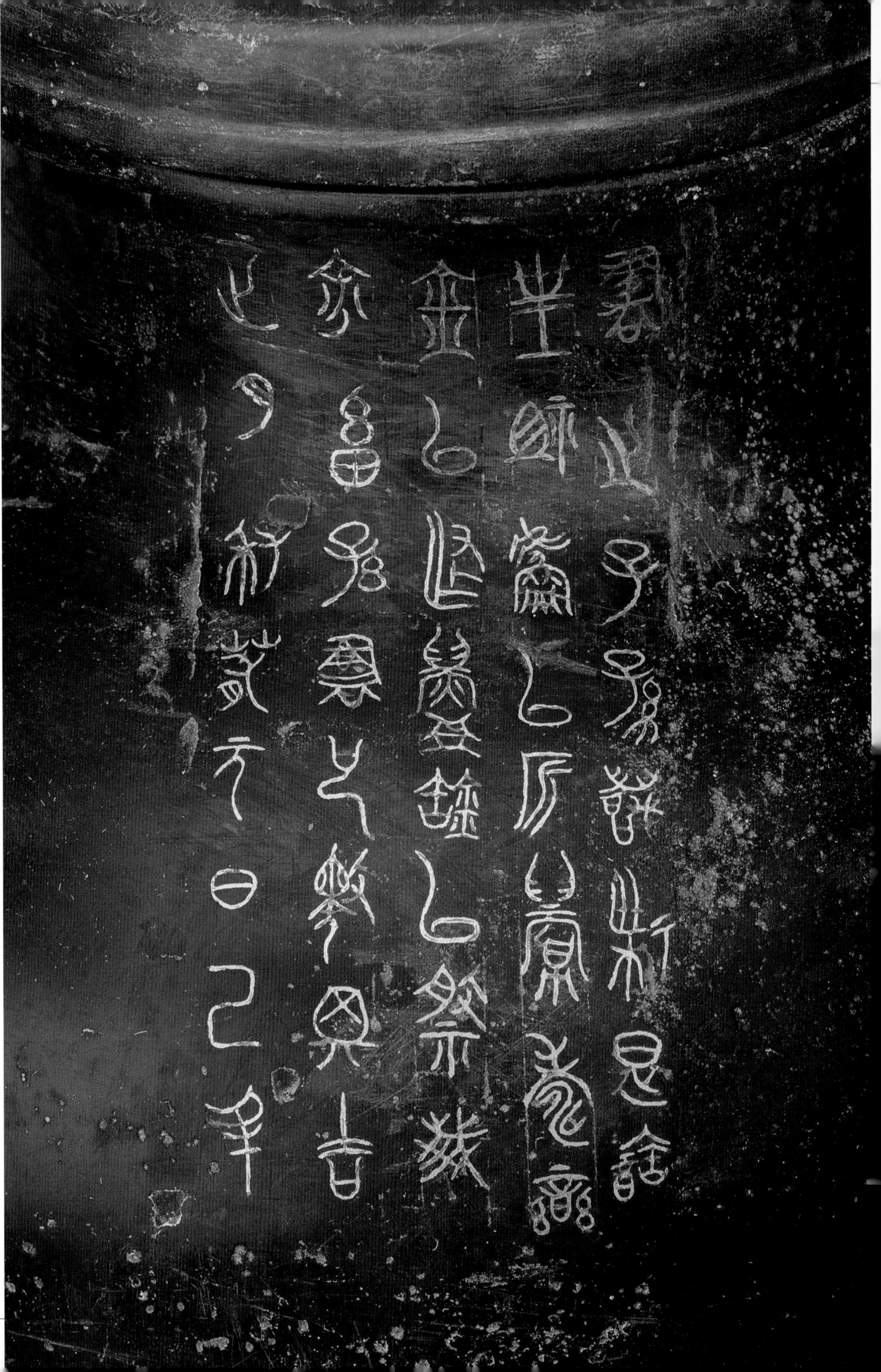

9-14*

9-14

9-14 Luan Shu fou, 40.5cm high, its mouth 16.5cm in diameter. The bottom has a diameter of 17cm. Belonging to the Spring and Autumn Period (770BC-476BC), it is preserved in the Museum of Chinese History. The inscriptions are inlaid in gold.

9-15 Copper zun with bird and animal design in gold plating, 14.6cm high, mouth 19.7cm in diameter and bottom 19.3cm in diameter. It is preserved in the Museum of Chinese History. Gold plating is an ancient technique, which began with the Warring States Period and thrived in the Western Han Dynasty. As it is difficult to oxidize gold, which is resistant to erosion, gold plating is often used to protect and decorate the surface of wares. This zun was made in 41BC, 3rd reigning year of Yongguang, Western Han Dynasty.

9-15

9-16

9-16 Shang Dynasty pot furnace for melting copper, 32.7cm high, diameter 30.5cm and bottom 3.5cm, unearthed in Erligang, Zhengzhou in 1953 and preserved in the National Museum of Chinese History. The Shang Dynasty (16th century BC-11th century BC) had many specifications of pot furnace to make alloys and answer the need of metal casting.

9-17 Shang Dynasty moulds for casting fang yi (rectangular casket shaped vessel with cover surmounted by knob), unearthed in the ruins of the Yin Dynasty, Anyang, Henan Province. It measures 25cmX11.2cmX4.8cm and 24.5cmX11cmX4.7cm. They are preserved in the Anyang Workstation of the Archaeology Institute of the Chinese Academy of Social Sciences. The discovery of the moulds provides evidence to enable us to understand the raw materials for making moulds, the mould manufacture method, the makeup of moulds and feeders for casting in the Shang Dynasty.

9-17

9-18

9-19

9-18 Spring and Autumn Period iron sword with gold handle and head, 38.8cmX4.4cm, unearthed in Shaanxian, Henan in 1957. It is kept in the National Museum of Chinese History. The Spring and Autumn Period (770BC-476BC) saw the manufacture of iron tools in China.

9-19 Warring States Period iron knife, 20.2cm long. The largest is 1.3cm wide. Unearthed in Guweicun, Huixian, Henan in 1950, it is preserved in the National Museum of Chinese History.

9-20

9-21

9-20 Warring States Period mould for casting iron, 28.3cmX10.3cm. Unearthed from Xinglong, Hebei in 1959, it is kept in the National Museum of Chinese History. The mould was made by pouring method. Once made it could be used repeatedly. Invented in the Warring States Period, the mould was used to make iron tools of certain specifications.

9-21 Warring States Period iron hoe, 24.4cmX7.8cm. The hole for installing handle is 5.2cm thick. Unearthed in Guweicun, Huixian, Henan in 1951, it is now kept in the National Museum of Chinese History. The shape of the hoe is rather similar to that of the modern hoe.

9-22

9-23

9-22 Rubbing from Eastern Han Dynasty wall painting carved on stone (partial view) showing bellows. Unearthed from Hongdaoyuan, Tengxian, Shandong in 1930 it shows the shape of bellows, the shape of metal wares and their use.

9-23 Model of blower made of leather mounted on wooden frame, with blast nozzle, etc. It has an entry and exit for currents of air and is operated by man. The model is made on the basis of the "Biography of Yang Xuan" in *History of the Later Han Dynasty*. Du Shi, prefect of Nanyang, improved the blower by using water to generate power to blow the furnace.

9-24 Han Dynasty piled-up mould, 29.5 cm X14 cm X 10.5 cm, unearthed from Wenxian, Henan in 1974. It is preserved in the Henan Provincial Cultural Relics and Archaeology Institute. The clay moulds were piled up layer after layer to allow raw materials to be packed together, using the same pouring exit and chamber which saved space and raised efficiency—a unique Chinese invention. It lowered cost and cut down liquid metal. The method began in the Spring and Autumn Period and reached maturity in the Han Dynasty.

9-25 Schematic diagram of piled up moulds.

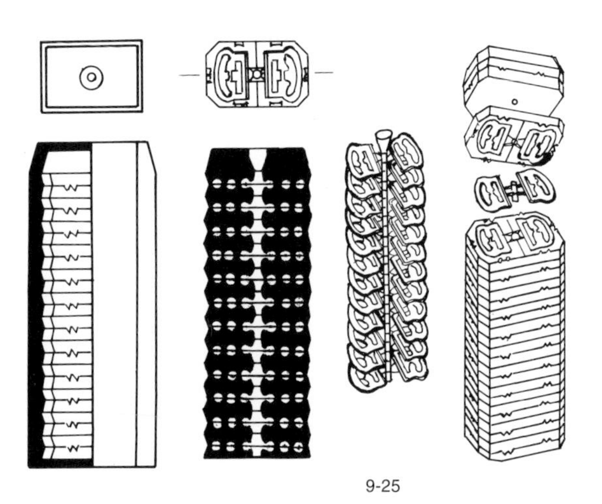

9-25

9-24

9-28

9-26 Complete mould (raised type or set in relief), 10cmX8.7cmX2.4cm, containing the characters "Da Quan Wu Shi, first edition". Belonging to the Han Dynasty (206BC-220AD), it is kept in the National Museum of Chinese History.

9-27 Eastern Han Dynasty iron knife (gold inlaid), 111.5cmX3cm, unearthed from Liuzhuangzhifangcun, Cangshan, Shandong in 1974. It is kept in the Museum of Chinese History. The texture of the knife is even, and there are 30 layers, which indicate that it had been heated and hammered 30 times. Having been decarbonated, it has become a very sharp knife. The words Yong Chu Liu Niang (112AD) can be seen on the knife.

9-28 Gold-inlaid inscription, Yong Chu Liu Niang, on the gold- inlaid iron knife.

9-26

9-27

Ancient Chinese Science and Technology

MACHINERY

One of the earliest countries in the world to have developed machinery, China made many inventions and mechanical installations. She is unique in the utilisation of energy.

In the field of simple mechanics, such as projectile, leverage, pulley, wheel and axle, she achieved considerable success in the early days. The Chinese made and used bow and arrow 10,000 years ago. Over 4,000 years ago they invented potter's wheel and manufactured fine porcelain. In the Xia Dynasty (2000B.C.) Xi Zhong made vehicles, using the principle of wheel and axle. Excavations of pits containing horse carriages in the Shang Dynasty ruins reveal wooden vehicles (some in square shape) that had one or two wheels. Wheel and axle were popularly used at the time. Chinese understanding of the wheel has a long history. In the Spring and Autumn Period jiegao was used—a pulley hung up on a tree to pull a bucket of water from the well by means of a rope. This reduced labour intensity of the early inhabitants of the land of ancient China. Through archaeological findings we see that scales were used in the Warring States Period. As a precision instrument, the weighing apparatus had been mastered early on. In the ruins of the New Stone Age, 6,000 years before our own time, we find many nets, which can cast open to catch fish by its own weight. Pointed bottoms of vases tell us that Chinese forebears utilised both the principle of gravity and centre of gravity to make the device.

In the Han Dynasty numerous machines came into use, based on a combination of principles of simple mechanics. Du Shi's shui pai or metal raft was a typical case of Chinese mechanics in those days. Du Shi used the wheel and axle to make water wheels to generate power in operating shui pai. He used the camshaft or wheel and shaft to blast the furnace through a leather chamber. A bamboo leverage was used as projectile to refill the air by means of rope in stretching open the leather chamber. This tells us the high level of advance in mechanics made by the people of the Han Dynasty.

Many achievements in utilising different kinds of energy were made by China. In addition to man power, domestic animals, water, wind and thermodynamics were applied as motive power. Over 3,000 years ago China used animals to draw heavy loads— horse to draw carriage or carts and ox to plough field. This was a direct way of using men or animals as a source of energy. The impact of rushing water to operate machines has a long history in China. The shui pai(using impact of water to push or operate) and shui dui (water-generated hammer to grind rice with the aid of a tanker) had been used in the Han Dynasty. The tongche or water driven wheel drew water from river to irrigate cropland with the aid of a bamboo tube, working 24 hours a day without respite. A labour saving device, it is used even to this day. Thermodynamics was applied to operate machines in the Song Dynasty. The horse lantern is the earliest example of the application of this principle in the world. The lantern burns candles, whose heat revolves the wheel, making the rider on horseback go round and round.

Wind power was discovered and put to use in China. Shang Dynasty inscriptions on tortoise backs and oracle bones carry the word fan (sail). The sail is a sheet of canvass, spread to catch the wind to propel a ship. By reason of its being carried through the air it is still used to drive windmills in salt fields in Southeast China to this day.

In the Spring and Autumn Period China successfully developed the pulley and windlass. The reeling machine was hoisted up and down. In early warfare defenders used pulley and rope to hoist a wooden box with men inside to watch enemy troop movement. The pulley was used to hoist vehicles up and down in mines during exploration. The device was unearthed from the ruins of an ancient copper mine in Tonglushan, Daye, Hubei Province.

The gearwheel was made in the same period in about the 7th century B.C. The mould for making gearwheel had been discovered in Houma, Shanxi Province.

Chinese knowledge and application of mechanical principles are deep and marvellous. Incense burner is an example of this knowhow. According to documents the incense burner, known as beizhong xianglu, was used in the Western Han Dynasty. The item was unearthed in Shapo village, Xi'an, Shaanxi. Made in the Tang Dynasty, the burner is shaped like a sphere in openwork, with two casings, homocentric ring, a center point and semispherical body. The supporting axle is arranged vertically from inner and outer rings and inner wall of the shell. Due to gravitation and vertical arrangement, the mouth of the silver incense burner always keeps upward no matter which way the burner turns. The joss stick stays inside the burner. The design is similar to the gyro-magnetic compass, which keeps itself in water level despite the tossing of ship in rough weather. It demonstrates the level of knowhow in mechanical principle of Chinese people of those days.

The manufacture of vehicles and ships occupies an important position in machine building industry. Vehicles were made by the Chinese in the Xia and Shang dynasties. War chariots were produced in large numbers in the Warring States Period. Vehicles underwent change in the Qin and Han dynasties as single shaft vehicles saw a reduction in output while two-shaft ones increased dramatically, with wide ranging varieties. The copper chariot known as safety vehicle on a single shaft, unearthed in the Tomb of Qin Shihuang (the First Qin Emperor), was drawn by a team of four horses. It has a sphercial cover on top. The Han Dynasty produced the big nianche, drawn by men. One type is with cover. Another has no cover. A copper nianche, unearthed in Xingyi, Guizhou Province and the pottery horse carriage, excavated in Chengdu, Sichuan, belong to this category. As vehicles of transport the Han Dynasty produced the single- wheeled cart, pushed by man. It carried cargo or passenger, convenient to operate in the plain or along narrow path in mountainous areas. In several hundred years in the run up to the Eastern Han Dynasty the ox cart became popular. A person can lie on it as it rumbles on rather steadily. One such cart unearthed from Caochangpo, Xi'an, is a typical cart belonging to the Northern Wei Dynasty. In late Palaeolithic Age Chinese forebears used the canoe. Wooden boats appeared in the Shang Dynasty at the latest. Wooden boats or ships appeared later in various structures, specifications and performances according to need. Chinese ships developed into a big family. Chinese inventions in ship building industry were numerous: steering apparatus, steering wheel and elevator and airtight cabins. The keel or long piece of wood along the bottom of a boat keeping the boat balanced in water is a Chinese invention, and so is the steering wheel, made ahead of other nations. The Chinese sail belongs to the hard type, which can be used to propel ship in all directions, whereas most other nations use soft sail. The generating power of ship depends on oars. Ships can sail with the aid of oars, which provide the generating power. That is why we see many oars in ancient ships. Shipbuilding technique reflects in a general way the level of development of science and technology, showing knowledge and application of fluid mechanics, mechanics of materials, resistance, leverage and relation between wheel and axle. In the Middle Ages China took the lead in shipbuilding technology. Of numerous ancient ships made in China, two deserve particular mention: the wheel- shaped oars that propeled boats on which were built cabins of the Tang Dynasty—forerunner of modern steamships. The treasure ship of Zheng He of the Ming Dynasty, 444ftX180ft, with 9 masts on which hang 12 sails was the biggest sailing boat in the world at the time. The big achievements in shipbuilding and navigation technology of ancient Chinese promoted economic and cultural exchanges between China and various countries and the development of sea transport.

10-1

10-1 Stone arrow head, 6.7cm long, unearthed from Sanliqiao, Shaanxian, Henan in 1957. It is kept in the National Museum of Chinese History. This New Stone Age arrow head shows that the Chinese had mastered the technique of making bows and arrows over 10,000 years ago. The appearance of the ballistic force machinery bolstered the power of man in extending the length of his arm to conquer nature.

10-2

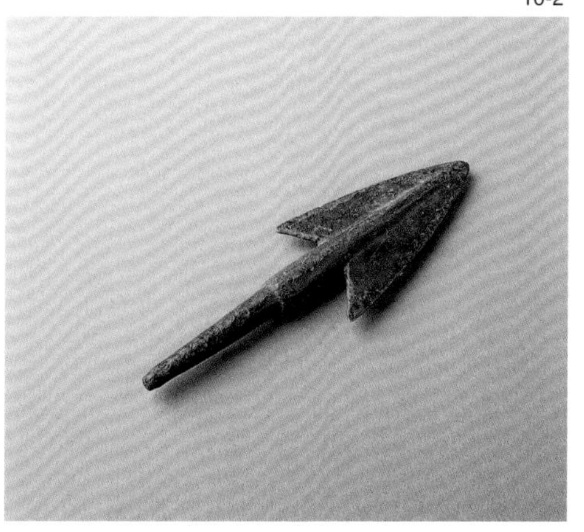

10-2 Shang Dynasty copper arrow head, 8.4cm long, preserved in the National Museum of Chinese History. Copper arrow head increased the offensive power of man in the Shang Dynasty (16th centuryBC-11th centuryBC). The arrow head was used mostly as a weapon during combat.

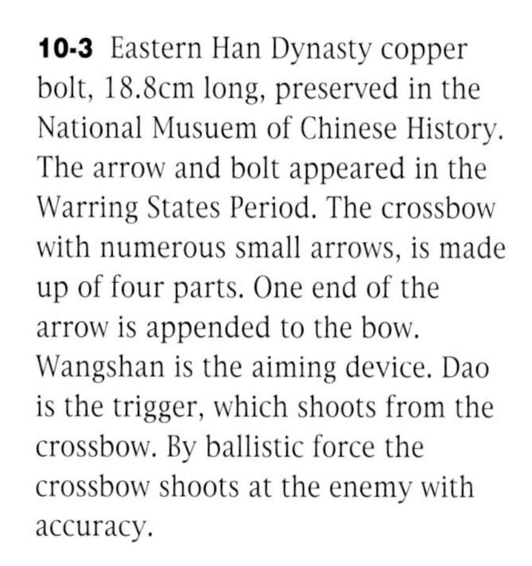

10-3 Eastern Han Dynasty copper bolt, 18.8cm long, preserved in the National Musuem of Chinese History. The arrow and bolt appeared in the Warring States Period. The crossbow with numerous small arrows, is made up of four parts. One end of the arrow is appended to the bow. Wangshan is the aiming device. Dao is the trigger, which shoots from the crossbow. By ballistic force the crossbow shoots at the enemy with accuracy.

10-4 Illustration of grave containing one-shaft horse carriage (two-wheeled) buried next to the tomb of Chinese aristocrat. The carriage was made in the Shang Dynasty. Knowledge of wheel and axle, the application of mechanical principles and manufacture of horse carriage became rather mature during the Shang Dynasty.

10-3

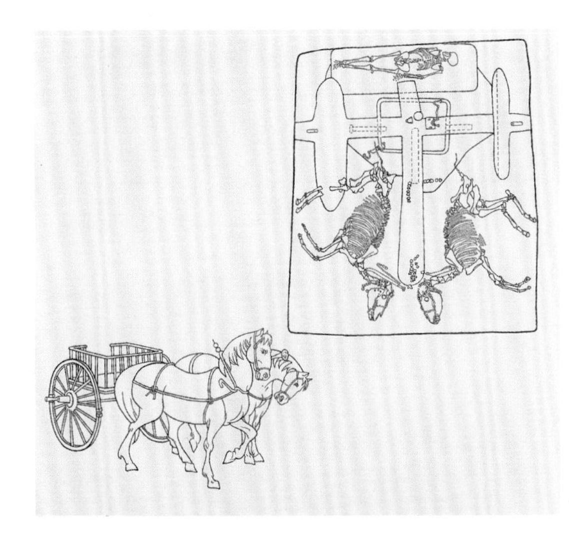

10-4

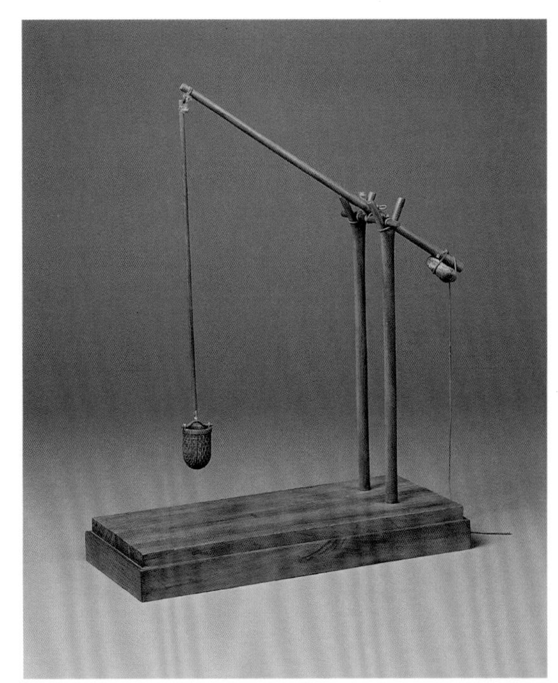

10-5

10-5 Model of the Spring and Autumn Period (770BC-476BC) jiegao. A person can hoist a bucket of water from well, using a stone on one side through a lever fixed on the tree. This is a device for hoisting water using lever principle.

10-6 Scales, 26.6cm long, 3.9-4cm in diameter, with metal units for finding weight, unearthed in Zuojiagongshan, Changsha, Hunan in 1954. They are preserved in the National Museum of Chinese History. They are made according to the lever principle. The metal units serving as weights are circular in shape. The scales belong to the the Warring States Period (475BC-221BC).

10-6

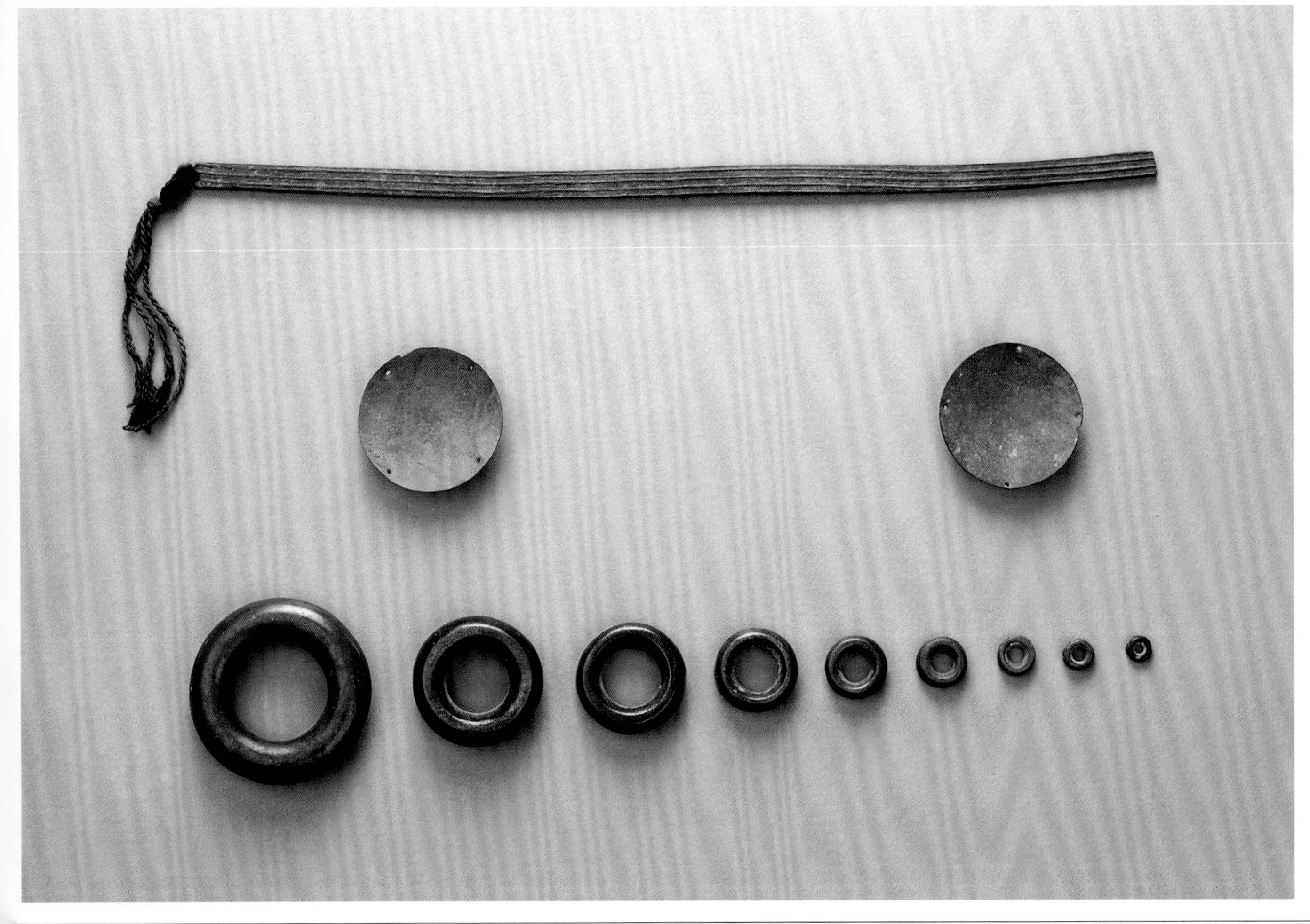

10-7

10-7 New Stone Age bottle with pointed bottom, 37.3cm high, 5.3 cm in diameter. Unearthed from Baoji, Shaanxi in 1958, it is preserved in the National Museum of Chinese History. Water can be drawn from the river with a rope, which is attached to the earlobe of the bottle. When the pointed bottom touches the water, the bottle will lean to one side. Water flows into the bottle through its mouth. When water reaches a certain level, the bottle will automatically stand up. The man pulls the rope and submerges the bottle into water until the bottle is full of water. This way of obtaining water is made possible by the application of the theory of centre of gravity and gravitational force by ancient Chinese forebears.

10-8 Metal shui pai can blast furnace with a lever and connecting rod to blow furnace. The wheel is water driven. The device was made by Du Shi, who utilised the impact of water, axle, convex axle, lever and ballistic lever to use water as power to blow air into the furnace in iron smelting.

10-8

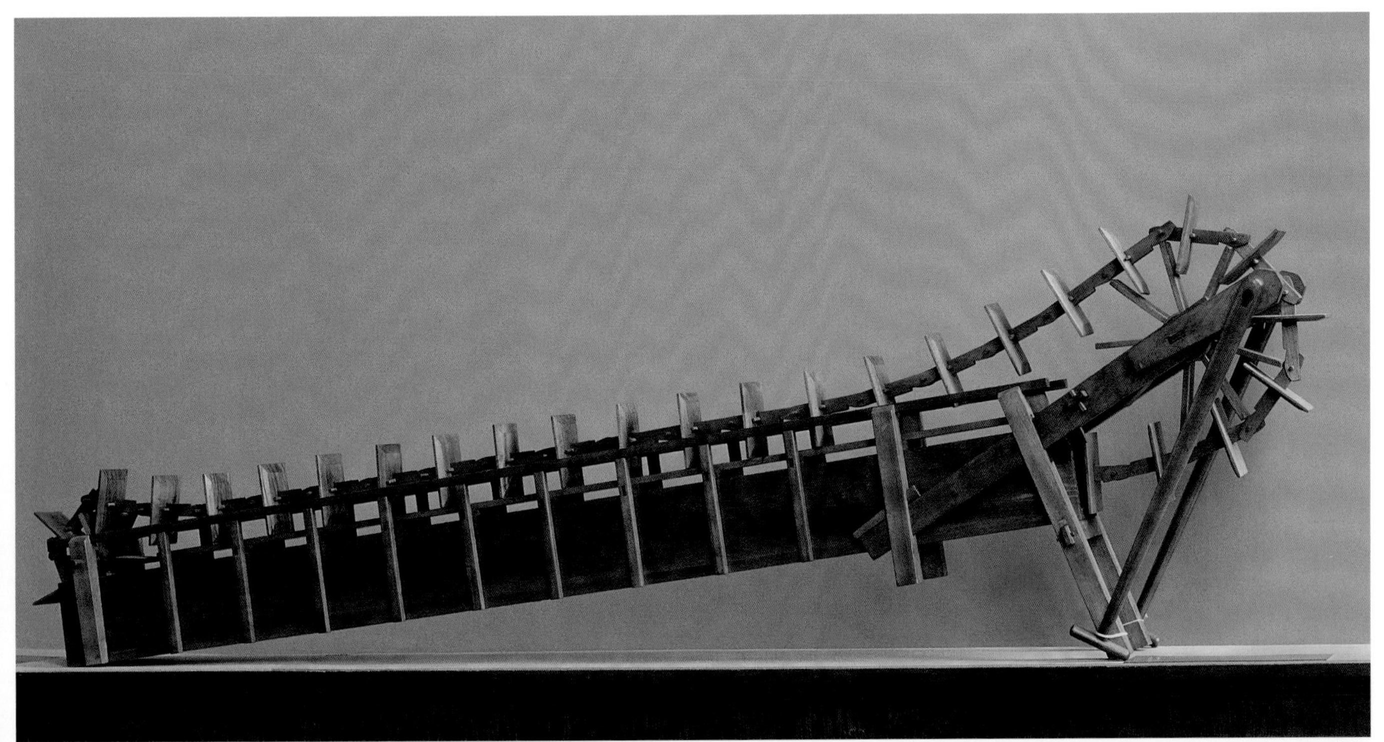

10-9

10-9 Model of a dragon-bone water wheel used for irrigating cropland. The device was invented by Bi Lan in the reign of Lindi, Eastern Han Dynasty (168—189). It was improved by Ma Jun in the Three Kingdoms, who promoted its use in the country. It is made up of handle, bent axle and gearwheel connecting rod. The device was operated by man, animal, water and wind power. Easy to make and rather efficient in hoisting water from river, it is used even to this day. The model is made on the basis of information obtained from the book: *Wei Lue* (Records of Wei) and traditional waterwheel.

10-10 Tongche was invented in the Tang Dynasty. It is made up of water wheel, bamboo tube, water trough,etc. It stands by the river. The water in the river pounds and turns the wheel, which enables the bamboo tube to carry water into the trough. Water irrigates cropland via the channel. As water in the river flows endlessly, the wheel turns endlessly too. This is an automatic irrigation device. The model is made on the basis of information from the book *Shui Lun Fu* (Ode to Waterwheel) by Chen Yanzhang of the Tang Dynasty and *Nong Shu* by Wang Zhen.

10-10

10-11

10-12

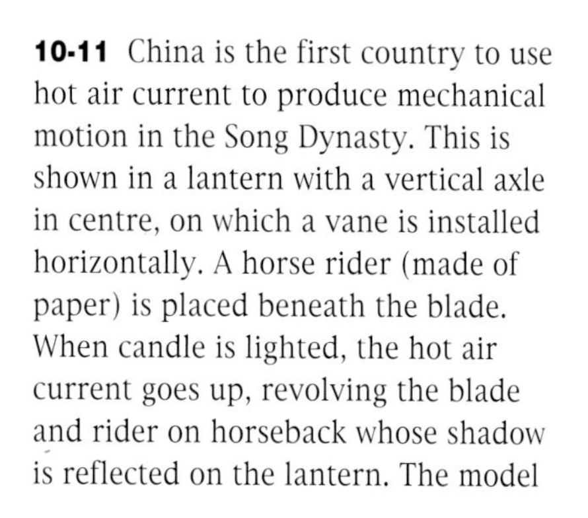

10-11 China is the first country to use hot air current to produce mechanical motion in the Song Dynasty. This is shown in a lantern with a vertical axle in centre, on which a vane is installed horizontally. A horse rider (made of paper) is placed beneath the blade. When candle is lighted, the hot air current goes up, revolving the blade and rider on horseback whose shadow is reflected on the lantern. The model is made on the basis of information obtained from the book: *Nan Hu Shi Ji: Shang Yuan Jie Wu Shi* (Collection of Nan Hu's Poems: Lantern Festival Poems) by Fan Chengda of the Southern Song Dynasty.

10-12 Han Dynasty pottery well, 47.8cm high, mouth 24cm long and 14.2cm wide. The diameter of bottom is 14.2cm. It was unearthed from Shaogou, Luoyang, Henan in 1952 and is preserved in the National Museum of Chinese History. The sliding gear was used—the first time ever in the 7th century BC in China. The hoisting or pulley tackle is evidence of the fact that sliding gear was used.

10-13

10-13 Western Han Dynasty copper gearwheels, 0.9cm high and 1.5cm in diameter, unearthed from Hongqingcun, Xi'an, Shaanxi in 1955. Of the same size the two objects are preserved in the National Museum of Chinese History.

10-14 This Eastern Han Dynasty vehicle records the number of wheel revolution. It beats a drum once the wheel travels one li in distance. This is

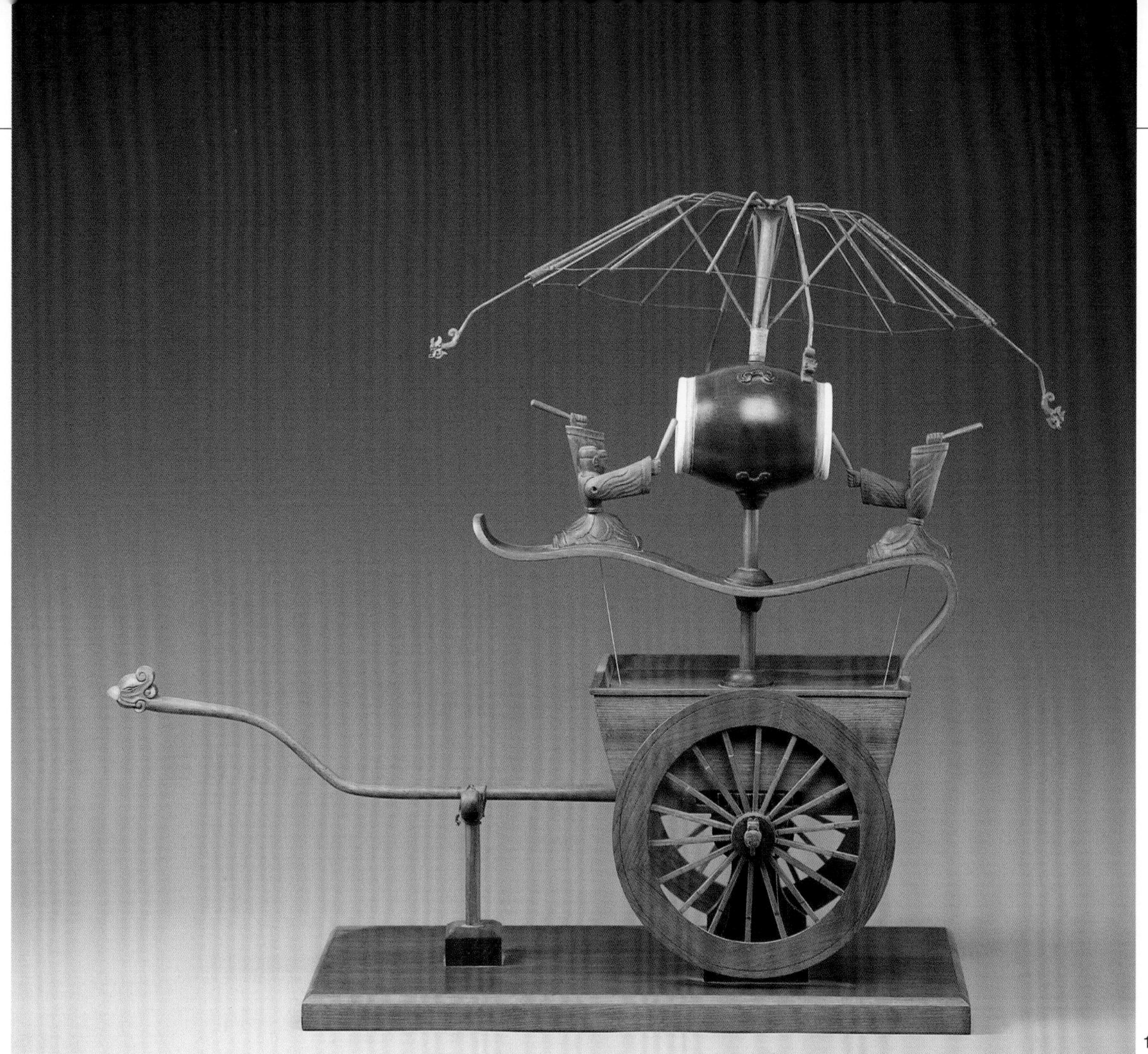

10-14

an ancient version of the modern taxi. The principles applied here are that when the wheel of the vehicle revolve so many times in covering one li it pushes the gear to revolve. This is due to lever principle of the convex wheel. The man beats the drum once for one li of distance. The model is made on the basis of information from the book *Song Shi: Yu Fu Zhi* (History of the Song Dynasty: Chapter on Vehicles and Dress) and drum vehicle from the wall painting in Xiao Tang Shan of the Eastern Han Dynasty.

10-15 Schematic diagram of the structure of the drum vehicle. The vehicle is made up of left and right wheels, a vertical gear wheel, a lower level wheel, a whirlwind wheel and mid level wheel. When the wheel of the vehicle revolves 100 times, the mid level wheel revolves once, which makes the wooden man beat the drum once to indicate distance travelled.

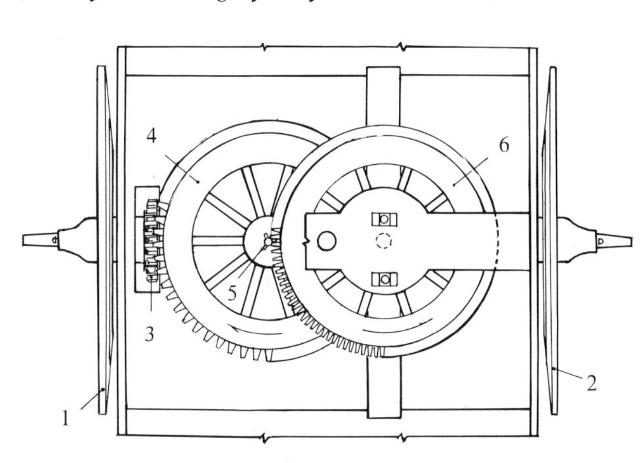

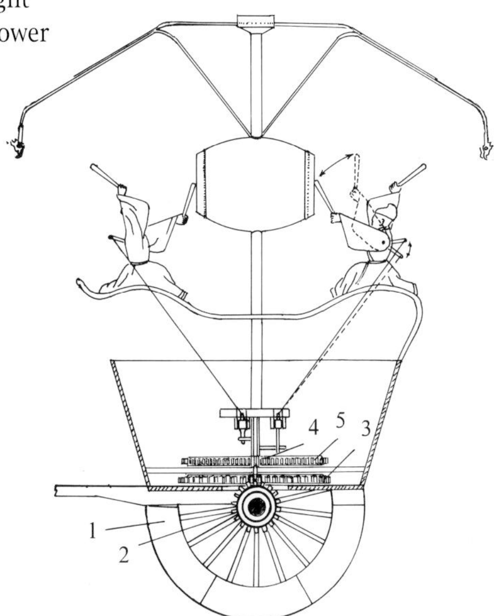

10-15

Structure of drum beating vehicle (a plane view)

1.left wheel 2.right wheel 3.vertical wheel 4.Lower plane wheel 5.whirl wind wheel 6.middle plane wheel.

Structure of drum beating vehicle (a side view)

1.right wheel 2.vertical wheel 3.lower plane wheel 4.whirlwind wheel 5.middle plane wheel.

10-16

10-16 Three Kingdoms compass vehicle, with gearwheel and clutch. Made by Ma Jun of the State of Wei, the vehicle may turn in different directions but the arm of the wooden man keeps pointing south. The model is made on the basis of information from the book *San Guo Zhi* (Annotations of the History of the Three Kingdoms) which gives quotations from *Wei Lue* and another book *Song Shi: Yu Fu Zhi.*

10-17 Schematic diagram showing the structure of the compass vehicle. The vehicle is made up of two wheels and a vertical gearwheel, small level wheel and big level wheel. During travel the left and right small level wheels are lifted up. The gearwheel remains inoperational. The direction of the wooden man remains unchanged. When the vehicle turns to the left, due to the swing of the shaft the left small level wheel is lifted up. The right small level wheel enters into position between the vertical wheel and big level wheel and becomes operational. When the vehicle turns left the wooden man turns right, the angle of turning is the same. So the wooden man keeps pointing south.

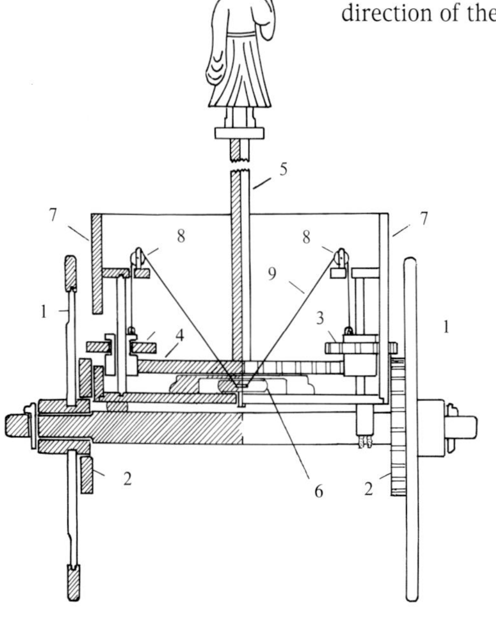

10-17

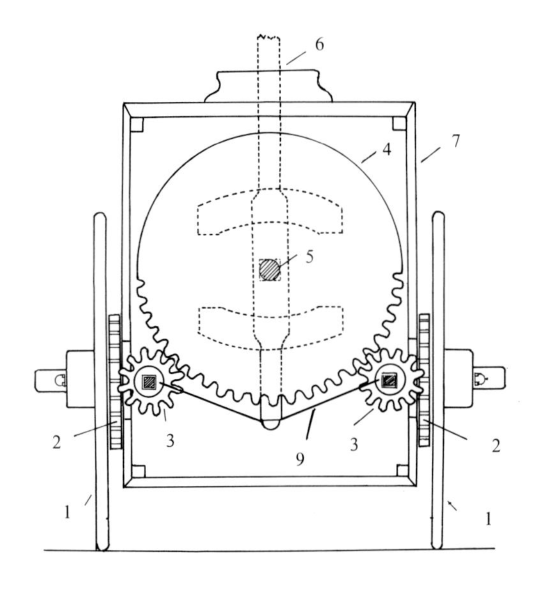

Hind view of compass vehicle

1.big wheel 2.vertical wheel 3.small plane wheel 4.central big plane wheel 5.cored vertical shaft 6.shaft 7.carriage 8.pulley 9.rope

Vertical view of compass vehicle

1.big wheel 2.vertical wheel 3.small plane wheel 4.central big plane wheel 5.cored vertical shaft 6.shaft 7.carriage 9.rope

10-18

10-18*

10-18 Tang Dynasty silver incense burner, with a diameter of 4.8cm at the belly. Unearthed from Shapo Village, Xi'an, Shaanxi in 1963, it is kept in the National Museum of Chinese History. It is made up of burner, inner and outer parts and outer shell. The ball bearings are arranged vertically. Regardless of how the burner stands the mouth of the burner always turns upright.

10-19 Copper gongche, 112cmX53cmX88cm. Unearthed from Xingyi, Guizhou, it belongs to the Eastern Han Dynasty (25-220AD). The carriage is kept in the Guizhou Provincial Museum.

10-19

10-20

10-20 Model of wheel-barrow. It was made in the Western Han Dynasty (206BC-8AD). Pushed by man, it is a convenient transport vehicle that can negotiate different mountain path or travel in the paddy fields despite its lack of steadiness. The model is made on the basis of a wall painting of the Eastern Han Dynasty.

10-21 Northern Wei Dynasty pottery ox carriage, 40cmX21cm X22.5cm. Unearthed from Caocangpo, Xi'an, Shaanxi in 1953, it is kept in the National Museum of Chinese History. Since the Han Dynasty Chinese aristocrats travelled on ox carriage. The form of transport was popular in the Wei-Jin era. The carriage is equipped with two shafts and arched canopy with windows on both sides. The passenger enters or exits by the back door.

10-21

10-22

10-22 Eastern Han Dynasty pottery flat boat, 56cmX15.5cmX16cm, unearthed from the Han Dynasty grave in the eastern suburb of Guangzhou city in 1955. It is kept in the National Museum of Chinese History. The boat has a flat bottom, flat bow and flat stem. There are three compartments. The rear compartment houses the quartermaster or helmsman. The helm resembles an oar. This shows that the helm was derived from the oar. There are three frameworks to place oars, which provide the generating power of the ship. The pottery ship reflects the actual shape of boats of the Han Dynasty.

10-23 Model of a flat-bottom passenger ship on the Bianhe River. It measures 86cm long, 62cm high and 25cm wide. The side of the ship is dovetailed by wood and reinforced with iron nails. A wooden pole stands in the middle of the ship to allow the ship to be towed by men. There is a wooden capstan at the bow to drop anchor. The stem has a steering wheel to balance the ship.

10-23

POSTSCRIPT

The *Artefacts of Ancient Chinese Science and Technology* has been compiled under the leadership of the State Cultural Relics Bureau, with the help of cultural relics units and museums at the provincial, municipal and autonomous regional level. Thanks are due to the following for their splendid co-operation: Archaeology Institute of the Chinese Academy of Social Sciences; China Printing Museum; Hebei Provincial Cultural Relics Bureau; Hebei Pronvincial Cultural Relics and Archaeology Institute; Shaanxi Provincial Cultural Relics Bureau; Shaanxi Provincial Cultural Relics and Archaeology Institute; Cultural Relics Bureau of the Xinjiang Uygur Autonomous Region; Xinjiang Uygur Autonomous Region Museum; Xinjiang Uygur Autonomous Region Cultural Relics and Archaeology Institute; Qinghai Provincial Department of Culture; Qinghai Provincial Cultural Relics and Archaeology Institute; Gansu Provincial Cultural Relics Bureau; Gansu Pronvincial Cultural Relics and Archaeology Institute; the Display Center of the Grottoes Cultural Relics Protection Institute of the Dunhuang Research Institute, Gansu Province; Cultural Relics and Archaeology Institute of the Ningxia Hui Autonomous Region; Helang County Cultural Bureau, Ningxia Hui Autonomous Region; Dalian City Cultural Relics Management Office; Lushun Museum; Jiangsu Provincial Cultural Relics Bureau, Xuzhou Museum, Jiangsu Province; Yangzhou City Museum, Jiangsu Province; Suzhou Silk Museum, Jiangsu Province; Anhui Provincial Cultural Relics Bureau; Nanling County Cultural Relics Management Institute, Anhui Province; Jiangsu Provincial Cultural Relics Bureau; De'an County Museum, Jiangxi Province; Shangdui Cultural Relics Management Office, Linchuan City, Jiangxi Province; Department of Culture, Hubei Province; Huangshi City Museum, Hubei Province; Jiangling Museum, Hubei Province; Ezhou City Museum, Hubei Province; Hunan Provincial Cultural Relics Bureau; Hunan Provincial Cultural Relics and Archaeology Institute; Fujian Provincial Department of Culture; Museum of Quanzhou Overseas Maritime Transport History; Guizhou Provincial Department of Culture; Guizhou Provincial Museum; Henan Provincial Cultural Relics Bureau; and Henan Provincial Cultural Relics and Archaeology Institute. Thanks are due to the following who have provided this album with the captions of the pictures: Feng Haozhang, Guo Qiying, Li Jianli, Wu Xingquan, Yu Zhiyong, Du Gencheng, Tu Junyong, Dao Erji, Li Qiang, Sun Hao, Wang Mingfang, Hu Guizhu, Xu Xinguo, Cui Zhaonian, He Shuangquan, Yan Weiqing, Lang Shude, Zhang Qingtao, Si Shengqing, Yang Xiaoxia, Liu Guangtang Zhou Yimin, Sun Huizhen, Wang Lilin, Liu Pingsheng, Hu Jingsheng, Zhou Diren, Xu Jianchang, Zhao Shunzhen, Wang Lianmao, Lin Deming, Ding Yuling, Qin Wensheng, Mao Wuying, Li Xiuping and Li Suting.

Thanks are due to Mr. Wang Guanzhuo who has been chiefly responsible for writing of the text with the all-out cooperation of the various departments of the National Museum of Chinese History.

DU YAOXI

Deputy Curator,
National Museum of Chinese History

图书在版编目（CIP）数据

中国古代科技文物：英文／中国历史博物馆编.
北京：朝华出版社，1998.7

ISBN 7-5054-0565-9

Ⅰ. 中…
Ⅱ. 中…
Ⅲ. 科学技术－历史文物－中国－图集
Ⅳ. K87-64

中国版本图书馆CIP数据核字（98）第03705号

中国古代科技文物（英文版）

《中国古代科技文物》编辑委员会编

朝华出版社出版

中国 北京 车公庄西路35号 邮政编码 100044

深圳雅昌彩色印刷有限公司制版印刷

中国 深圳 上步工业区304栋西二层 邮政编码 518030

中国国际图书贸易总公司发行

中国 北京 车公庄西路35号
北京邮政信箱399号 邮政编码 100044

新华书店经销

1998年第一版 第一次印刷
ISBN 7-5054-0565-9/J·0287
08600
85-E-508P

中华人民共和国印制